Praise f
Earth's Wild Music

"*Earth's Wild Music* is a lamentation, an exaltation, an impassioned indictment, and most definitely a call to action."
—BARBARA LLOYD McMICHAEL, *Our Coast Weekend*

"Steeped in nature, brewing alternately with love and rage . . . Moore writes at the intersection of ode and alarm inhabited by the spirits of Mary Oliver and Rachel Carson."
—LENORA TODARO, *Catapult*

"[Moore's] is a thoughtful, insistent, necessary voice in the ongoing conversations about how to treat the natural world."
—KATIE NOAH GIBSON, *Shelf Awareness*

"Taken together, the essays demonstrate the many ways in which a nature lover can celebrate and advocate for the beauty of Earth, even as it faces widespread, human-caused destruction."
—AMY BRADY, *Literary Hub*

"Exceedingly knowledgeable, experienced, and expressive . . . Moore details all that we're losing to climate change, spiking gorgeously precise descriptions and dramatic tales of wildlife encounters with grim statistics about the escalating die-off of birds and other species, the 'great starving' underway in the oceans, and the ongoing destruction of forests and wetlands."
—*Booklist* (starred review)

"An enthusiastic argument that love, care, and defiance may still save the Earth."
—*Kirkus Reviews*

"Kathleen Moore is an ambassador for listening, and *Earth's Wild Music* is a guidebook for how this untapped sense can lead us to falling in love with our true home: the world of nature." —PAUL WINTER, saxophonist, composer of "Wolf Eyes"

"*Earth's Wild Music* is both a poignant meditation on extinction and a stirring call to resistance; it brings the radiant songs of the wild world within hearing." —JULIAN HOFFMAN, author of *Irreplaceable: The Fight to Save Our Wild Places*

"An enchanting book, *Earth's Wild Music* draws the reader into the sounds of the natural world and provides inspiration to reverse their decline. Kathleen Dean Moore has a very special and authentic voice. A compelling and magical read." —THOMAS E. LOVEJOY, Senior Fellow at the United Nations Foundation and professor of environmental science and policy at George Mason University

"[Moore] brings her fierce attention and generosity of spirit to the varieties of kinship that enjoy Earth's astonishing soundtrack. Loon, wolf, meadowlark, bear, canyon wren, Beethoven, Puccini, and Leonard Bernstein all assist her in this paean and plea for awakening the moral imagination and getting real about facing up to and transforming human destructiveness." —ALISON HAWTHORNE DEMING, author of *Zoologies: On Animals and the Human Spirit*

"Passionate and poetic, Kathleen Dean Moore creates a splendid ode to the biophonies extant in a shape-shifting world." —BERNIE KRAUSE, author of *The Great Animal Orchestra: Finding the Origins of Music in the World's Wild Places*

"In this poetic tribute to the beauty of the earth and its inhabitants, Kathleen Dean Moore brings her readers in touch with the spiritual way that birds and mammals, other animals, and natural sounds demonstrate our need to preserve ecosystems. A lovely read."

—PETER H. RAVEN, President Emeritus,
Missouri Botanical Garden

Earth's Wild Music

EARTH'S WILD MUSIC

Celebrating and Defending the Songs
of the Natural World

NEW AND SELECTED ESSAYS

Kathleen Dean Moore

COUNTERPOINT
Berkeley, California

The Library of Congress has cataloged the hardcover edition as follows:
Names: Moore, Kathleen Dean, author.
Title: Earth's wild music : celebrating and defending the songs of the natural world / Kathleen Dean Moore.
Description: First hardcover edition. | Berkeley, California : Counterpoint Press, 2021.
Identifiers: LCCN 2020017228 | ISBN 9781640093676 (hardcover) | ISBN 9781640093683 (ebook)
Subjects: LCSH: Extinction (Biology) | Nature sounds. | Philosophy of nature.
Classification: LCC QH78 .M66 2021 | DDC 576.8/4—dc23
LC record available at https://lccn.loc.gov/2020017228

Paperback ISBN: 978-1-64009-530-4

Cover design by Dana Li
Book design by Jordan Koluch

COUNTERPOINT
2560 Ninth Street, Suite 318
Berkeley, CA 94710
www.counterpointpress.com

Printed in the United States of America

10 9 8 7 6 5 4 3 2 1

For Frank,
who knows the world by the songs it sings.

Someone will say: you care about birds. Why not worry about people? I worry about *both* birds and people. We are in the world and part of it, and we are destroying everything because we are destroying ourselves spiritually, morally, and in every way. It is all part of the same sickness.

—THOMAS MERTON

Contents

Preface / The Work of Loving the World   xv

Prologue / The Music in Their Bones   3

1. *Tremble*

Symphony No. 9, Scored for Cactus   11
Earth's Wild Music   18
The Sound of Human Longing   23
Praise Songs I–VII (When Darkness Turns
    Unexpectedly to Light)   27
The Love Child of Father Time and Mother Earth   34
Repeat the Sounding Joy   43
Songs in the Night   53
Listening for Bears   61

2. *Weep*

The Tadpole Motet   69
The Silence of the Humpback Whale   75
The Meadowlark's Broken Song   85

The Terrible Silence of the Empty Sky (Intergalactic Space)   92
The Terrible Silence of the Empty Sky (Forest)   99
The Terrible Silence of the Empty Sky (Seashore)   105
Twelve Heartbreaking Sounds That Will Remain   113
Sorrow Fired to the Strength of Stone   115

3. Awaken

Living Like Birds   121
Sleep, Judy Francine   130
Alarm Calls   137
Another Marshland Elegy   142
Late at Night, Listening   150
Silence Like Scouring Sand   158
The Song of the Canyon Wren   166
How Can I Keep from Singing?   171

4. Sing Out

After the Fire, Silence and a Raven   183
We Will Emerge Full-Throated from the Dark Shelter
    of Our Despair (The Dawn Chorus)   191
The Sound of Mountains Melting   196
Rachel's Wood Pewee (On Wonder)   204
Hear the Wind Blow   212
Be the Bear   218
Hope Is Not the Thing with Feathers (A History)   228
Why We Won't Quit   235

Epilogue / Sing Out from the Mountaintops   239

Acknowledgments / A Chorus of Friends   245
Notes   247

Preface

THE WORK OF LOVING THE WORLD

FROM THE TIME I BEGAN WRITING ESSAYS, I HAVE CALLED MYself a "nature writer," although I have not always been sure what that sort of work entails. The poet Mary Oliver wrote, "My work is loving the world . . . which is mostly standing still and learning to be astonished . . . which is mostly rejoicing . . . which is gratitude . . ." and that seemed about right. Because I was so in love with the world, writing the love songs was simple and straightforward. You just go to some place wonderful, open your heart and your notebook, and tell the truth.

But then? After a time, loving the world became more complicated, and rejoicing got harder. Even as I was celebrating this splendid world, it was slipping away. I was midway through an essay on frog song when developers bulldozed the frog marsh for condominiums. I had just published an essay about an owl's nest in a favorite lodgepole pine forest when the forest, and the nest, burned to ashes and spars. As I celebrated their songs, humpback whales grew thin, starving in a warming, souring ocean. And all the while, executives of multinational extractive industries were gathering around mahogany tables to devise business plans that they knew would take down the great systems that sustain human life and all the other lives on Earth. Oh, the peril. The ecological peril. The moral peril.

In the fifty years that I have been writing about nature, roughly 60 percent of all individual mammals have been erased from the face of the Earth. The total population of North American birds, the red-winged blackbirds and robins, has been cut by a third. Half of grassland birds have been lost. Butterflies and moths have declined by similar percentages. As individual numbers decrease, species are being lost too. As many as one out of five species of organisms may be on the verge of extinction now, and twice that number could be lost by the end of the century. Two-thirds of the species of primates, our closest relatives, are endangered. Unless the world acts to stop extinctions, I will write my last nature essay on a planet that is less than half as song-graced and life-drenched as the one where I began to write. My grandchildren will tear out half the pages in their field guides. They won't need them.

The loss of species scares me. The loss of their music breaks my heart. Each time a creature dies, a song dies. Every time a species goes extinct, its songs die forever. How will we live under the terrible silence of the empty sky? My nightmare is that before we lose the Earth's life-sustaining systems, we will lose its soul-sustaining system—the Earth's wild music—and all that will be left will be the immortal Dolly Parton and methane burps.

Now, in the shadow of the Sixth Great Extinction, a pandemic has spread a terrible new silence over the world. What does a nature writer do? What does she write?

One ordinary temptation is to give up and become something other than a nature writer, turning to polemics, ethical treatises, signboards, and lots of nasty letters. But how can a writer leave the leafy world for the marble halls of politics? Another temptation is to stop writing altogether. "If you don't have something nice to say," my mother insisted, "then don't say anything at all." But how could she ever have anticipated the drenching grief in the stories now asking to be told? The temptation in ordinary times would be simply to lie by omission, continuing to tell only the happy stories, ignoring the difficult fact that the world is being assaulted and ravaged at exponentially accelerating rates.

But we do not have the luxury of writing in ordinary times. I'm convinced that writers are therefore called to efforts that are out of the ordinary. "We need the books that affect us like a disaster, that grieve us deeply," novelist Franz Kafka wrote. "A book must be the axe for the frozen sea inside us."

So. How exactly does a disaster affect us? We know the answer from experience, having been cruelly tutored by world wars, hurricanes, and a pandemic. Disasters call us to action. They call us to levels of compassion and courage we did not know we could reach. They smash us with sorrow and lift us with determination and moral resolve, the way a wave both smashes and lifts us in the same wild movement. Disaster transforms sorrowful love into a force strong enough to change the trajectory of history. I don't know if any book can do what Kafka asked. But dear Mary Oliver, do you think this might now be how we do the work of loving a weary, reeling world? And don't we have to try?

In *Earth's Wild Music: Celebrating and Defending the Songs of the Natural World*, I tell stories about the planet's imperiled music, one consequence of our civilization's having lost its way. Why music? Because I love it. I often tell audiences what Frederick Buechner wrote: that "you will find your calling at the intersection of your deep love and the world's deep need." This is surely true for the nature writer. So I write from that place where my deep love for the world's music—the birdsong, the frog song, the crickets and toads, the whales and wolves, even old hymns and Girl Scout songs—meets the terrible facts of onrushing extinction.

MY NEIGHBOR TO THE WEST WAS CONFUSED WHEN I SAID I WAS writing about the extinction of Earth's wild music. "Wait," she said. "Is that a *thing?*" Dear god, yes. Extinction of the world's species is in fact a thing, and with the species will go their songs. To the extent that people don't know this, it is the responsibility of the nature writer to tell them: to bear witness, ring the church bell, trip the alarm, beat the warning drum, send the telegram, blow the whistle, call all-hands-on-deck—and sometimes, weeping, to write the condolence letters.

MY NEIGHBOR TO THE NORTH WAS WORRIED. "WAIT," SHE SAID. "Why would anybody read such a sad book?" We need to talk about the need to face sorrow straight-on, I tell my neighbor. Oh, she knows grief. It has not been her friend. She has more than once followed the five stages of grief, traditionally listed as Anger, Denial, Bargaining, Depression, and ultimately Acceptance.

In my experience, the extinction and climate-change catastrophes have led the world's defenders through the five stages of grief to arrive at the worst possible place. At first, we seethed with fury, but what enemy could we identify, back then? Then, we refused to believe that we stood to lose it all, but the evidence was sound and only the ideologically ignorant are still in denial. We tried to bargain with the forces of destruction, but when a CEO is making a killing, why would he settle for making a living? We fell into depression. And now, who among us is not fighting with every ounce of her strength to resist falling into acceptance, abdicating all resistance and be-lieving that there is no hope of recovering a singing, surging, life-sustaining planet? But acceptance is defeat. It is morally impossible.

So this book draws a new map through sorrow to something more powerful.

Part One. The essays begin with *Tremble.* We tremble with joy and won-der when we open our hearts to the music that comes to us out of thin air, the astonishing and mysterious gift of the tremulous universe.

Part Two. Then, as nature's voices fall away, the essays call us to *Weep.*

The silence of morning and marsh is unbearably lonely and sad. How will the children dance without wild music? How will they live without rustling birch trees and chickadees?

Part Three. Weeping is the start of grieving, not the end. It invites— no, it morally requires—movement toward *Awaken*. Our grief is not only a measure of our love but a measure of our obligation. It is therefore our responsibility to awake to the work of saving what we can of the songs.

Part Four. *Sing Out*. I can't think of any other morally possible place to go. By resolute effort, maybe we can save some of the endangered beings. And if we can't, then by that effort, we perhaps will save some of our moral integrity, which is consistency between what we believe is right and the actions we choose to take.

Epilogue. When the book is over, we will talk about what we can do and how we will do it.

MY NEIGHBOR TO THE EAST, A WRITER HIMSELF, WANTED TO know how I was going to compose the book: contents, characters, setting, etc.

I explained that the thirty-two essays in the book are drawn from a lifetime of loving the world. The majority are new essays written in response to the extinction crisis. When I felt the book needed celebration, I pulled in familiar essays, or parts of them, that I had published in early books, when loving the world seemed pure and simple. Some of the essays are mosaics that gather fragments of old and new essays and arrange them to portray something unforeseen.

The main character is my husband Frank, to whom the book is dedicated. He is a tall, square-jawed guy, handy around tools and boats, as Alaskans must be, strong and smart—but you'll see. He is also a behavioral neuroendocrinologist, retired from Oregon State University after having studied, in the rough-skinned newt, how external cues, like the presence of a potential sex partner, register in the chemicals of the brain and change the behavior of males.

The book is set in a variety of the seasons and places we have traveled in Minnesota, Manitoba, California, Arizona, and especially Oregon, which is our home. But most of the essays are set in the place where my husband and I spend the summers, a little cabin on a hill where two creeks and a bear trail meet a tidal cove on one of Alaska's ABC Islands. We keep a small fishing boat there and a couple of kayaks, so we can range up and down the inlet. A waterwheel about the size of a birthday cake powers light bulbs and our laptops, unless the stream level falls and the hydropower fails, which is why we carry flashlights and ballpoint pens during droughts.

But right now, I am at the desk that looks into our tiny backyard, here on College Hill in Corvallis. A gale poured over the Coast Range this morning, so the Douglas-fir limbs are luffing in unpredictable winds. The chickadees and juncos have fled the birdfeeders as evening comes on, and a gray squirrel is now pirating the feed. Rain pounding on the deck is the music I hear. The garden is green and wet and primevally alive. It's the end of a long day of loving the world.

KDM
Corvallis, Oregon

Earth's Wild Music

Prologue

The Music in Their Bones

IT'S NOT EASY TO LAND A BOAT ON THIS WILD ISLAND. IT SITS alone in the mouth of a Southeast Alaska bay, smacked by tidal currents that race up and down the channel. The island's basalt cliffs rise twenty feet from the water, buttressed by fallen boulders, topped by a woven palisade of hemlock and Sitka spruce. But on the south curve of the island, the cliffs collapse into a pile of boulders and shelving stone. In calm weather, it's possible to nose a skiff into these rocks and jump onto a submerged ledge. From there, it's a haul and scramble between boulders.

We couldn't leave the boat tied on a falling tide. Just one contrary wave and the bow would hang up on rocks and stay pinned there, tilting, until the returning tide lifted it again. So my husband landed us gingerly, our two grandchildren and me, and motored into the bay to hold the boat.

On the island, the boulders were the size of woodsheds. The best route seemed to be between them on blue paths of broken mussel shells, and under their overhangs in the dank world of starfish, and over them, up the splintered wave-tossed logs, and around them in water to the top of our boots. It was a long scramble for the little boys. But as soon as we entered the rift in the forest, we found a narrow trail blanketed in moss. It was too delicate a path to have been made by boots—a deer trail instead, I thought, or a trail made by bears. So I was listening carefully, stopping often to hear what we might disturb. What we disturbed was a raven on its nest. It screeched, flapped over, and dropped a stick in our path. A present, a child said. A warning, I thought.

Although trees barricaded the island's edges, inside that wall was an open nave. Lifted on the black pillars of hemlocks, a green ceiling vaulted high overhead. In a column of light, red columbine nodded and orchids bloomed. Not the big, birdlike orchids of the tropics, but northwest orchids, a pallid spiral of tiny shoes on a stem. We found tunnels under tree roots in the sanctuary, lots of them, dark pathways beaten smooth by small paws. The mossy carpet was clean. But not just clean. It had been swept of all its litter. The hemlock cones, sticks and twigs, ruffled lichen, needles, and branch tips that usually litter a forest floor had all been gathered into small piles, as if someone were expected to come along with a dust pan to sweep them up. If I hadn't recognized these as the scent posts of river otters, I would have imagined small nuns singing softly as they glided across the transept with little brooms—the space was that perfectly cared for.

In the nave, Swainson's thrushes filled the air with flute-song. Although the boys are taught to sing out for bears, they were quiet here, listening. Light flowed between branches, and because the branches were gently swaying, the light swayed too, and reflections played up from the sea.

There, splayed on a mossy bank as if on an altar, was the skeleton of a bald eagle. A large one, so most likely a female, gone mostly to bones. She lay on her back with her wings spread, looking toward the east. Only a few magnificent primaries, the feathers of flight, stuck to the reaching wingbones. There was the jutting keel of her breast, and a cage of ribs above scattered vertebrae. The eagle's long leg bones were dull and half buried in moss. The talons remained, although torn somehow off to the side, as if she had dug dirt in the agony of dying. At the top of the spine was her skull, staring with empty eye sockets along the crest of her ferocious beak.

Whoa, a grandson whispered. But I, too, could not have been more surprised. In retrospect, I don't know why. If we were ever to come upon the bare bones of an eagle, fallen somehow on her back as if blown from a high perch, or laid as if in a sacrificial ceremony on a bed of moss, this is where we would have found her, in the dim light of this wild island. We circled the eagle, touched her. The little boys crouched beside her and stroked her beak. Then I lay on my back beside the skeleton. I wanted to see what the eagle must have seen as she died, to feel the moss on my back and the damp light on my face. I lay there a long time, looking east, listening to the wind and the water-smack. Thinking.

I didn't know what to think, there in the presence of this majestic death and two small, eager lives. So much loss surrounds us. I, and all people of my generation, were born into a world packed with life and beauty. Without much noticing, we have lived through the destruction of almost half of it, plowed or burned, poisoned or killed, transmogrified into products or into human flesh, leaving the world half stripped to its rocky bones. So much peril surrounds the children, who were born to this great planetary decision point, soon to witness the rapid reinvention of human civilization or its slow extinction.

Extinction. Extinguish. Cause to cease burning—all the little lives, all the small songs.

I can't help but think of dinosaurs, the progenitors of the eagles and of all birds. I imagine a tiny theropod breaking open her leathery shell. She sticks out her scrawny neck and emerges into a sea of sound, the rustle-squeak and drone of the crowded forest. It's even possible that in the sky, she sees the golden streak of the asteroid that will smash into the Earth, the beginning of the Fifth Extinction. The theropod would have gasped—maybe baby dinosaurs' only expression is gasping—not at the meaning of the streak, surely. But how could she not gasp at the chorus of sound? How could she not flinch at the sudden light?

My grandsons were born into a world powered by fossil fuels on fire, the beginning of the Sixth Extinction, the end for uncounted species on the planet, and perhaps their own species. This last possibility is a matter of serious, hideous debate. What is beyond debate is Paul Ehrlich's warning: "Few problems are less recognized, but more important than, the accelerating disappearance of the earth's biological [richness]. In pushing other species to extinction, humanity is busy sawing off the limb on which it is perched." Closing their eyes, those baby boys curled their toes and melted into the murmurs of their mother, who loves them more than life. So there we are.

It's hard to know what to think and how to live in the context of the paradox and peril that face humans and all the other animals.

Our duty at this hinge point in history, some say, is to be grateful and glad. Our role, some say, is to celebrate the Earth and to love it. Our challenge, my friend Hank says, is to "find new beauty in the rushing changes." The moral obligation of those who love life itself, some say, is to be still and rejoice in the music of the singing world. The world is still beautiful; celebrate that. It's true: This island is glorious. The eagle's bones are light-swept under a fine lace of lichen. The thrush song is woven from shining threads of happiness. The children are a delight ongoing. Already, they are wondering how to use their pocketknives to gouge a flute from a deer femur they found on the beach.

It can be done, you know. For forty thousand years, people have shaped flutes from the femurs of dead birds. Neanderthal artists, before they vanished, carved whistles from the bones of baby cave bears, now also vanished. In a Utah canyon, I have heard the music of a flute made from the leg bone of an eagle. Not at all the skitter-scrape of a living eagle, that flute made a lovely, lonely sound, the music of wind through bleached bones. *All things shall perish from under the sky. Music alone shall live.*

But tell me: What is the song of a dead bird? What is the whistle of vanished wings? What music does the mocking rain play on the suddenly mute island?

I will howl against the approaching silence of the empty sky. I will carve a flute from a dead bird's bone and whistle like a bosun on a sinking ship. I will accept sorrow as a last great offering from a desperate world. But then I will shape anguish into something that is fierce enough to stand in defense of all we love too much to lose.

I know that whatever is left of the planet when the pillage ends, that's the world that the children will live in. Whatever genetic song lines, whatever fragments of whale squeal and shattered harmonies are left, that's what evolution has got to work with. Music is the trembling urgency and exuberance of life ongoing. Truly, if we can't save the songs, can we save ourselves? In a time of terrible silencing, what can we hear if we listen carefully, and what can Earth's wild music tell us about how we ought to live?

1.

TREMBLE

LISTEN: A THUNDERSTORM HAS MOVED OVER THE MESA. RAIN-drops strike the desert hardpan. Thunder shakes the air. Under the sand, an estivating spadefoot toad feels the vibrations that signal rain. He opens his bulging gold eyes. With spades on his limbs, he disinters himself, emerges into the steamy, soaked world, and croaks. It's a metallic sound, like a spoon dragged over a cheese grater.

Sound is vibration. Music is sound organized in time. The universe trembles with wild music. Who would not tremble with wonder?

Symphony No. 9, Scored for Cactus

IN THE DAYS SINCE WE CAME HOME FROM THE SONORAN DESERT, I have been hanging out on YouTube, watching Leonard Bernstein conduct the Vienna Philharmonic Orchestra in Beethoven's Symphony no. 9, Movement 4—the "Ode to Joy." The maestro pouts his lower lip, swipes the hair out of his eyes, and off they go. Pulling music from the air, Bernstein throws it at the flutes, then the kettledrums. Then he reaches his arms wide and lifts his head as if he is conducting a meteor shower. The massed choir and orchestra respond perfectly, blazing through space, screeching and resounding. The "Ode to Joy" moves me to tears every time I hear it—the greatest piece of music ever composed, I believe, "a vision of joy in the temple of nature," as Beethoven's biographer wrote.

LAST WEEK MY HUSBAND AND I WERE CAMPED AT THE MOUTH OF a desert canyon that gave onto an alluvial fan studded with organ-pipe cactus, ocotillos, and the magnificent saguaro cactus, all in bloom. Frank says saguaros look like outlaws with their arms raised in surrender. To me, they look like women in narrow pleated skirts, offering white flowers to the hummingbirds and long-nosed bats. But the true gifts of the saguaro are the stiff spines set in clusters on the pleats of their trunks. When the wind blows across the spines, they sing like violin strings. Better yet, when you pluck a spine, it will sing its particular tone. If a person is patient in her plucking, she can play music on a saguaro cactus.

Frank had wandered off to look for a Gila woodpecker that was rasping in the draw, so I found a good saguaro and set about plucking and listening, plucking and listening, until I had figured out which eight spines played a major scale. Now, what to play that would be worthy of the magnificence of the day? "Baa, Baa, Black Sheep" was within my skill set. "Twinkle, Twinkle, Little Star"? In the end, I played the simple theme of the "Ode to Joy," repeating it, the way the tenors repeat it, and the cellos.

But then the wind came up, as it often does toward the end of the day, and I stopped my plucking to listen. Scuffling through mesquite leaves, a collared lizard scurried under a rock. Crickets began to saw away with their violin-bow wings. Lesser nighthawks rolled their *r*'s like great Italian tenors. Above the small noises, the shared voices of all the spines on the saguaros sang out.

This was the music that the desert played under the falling sun. It was whispered and warm—no brass, all strings. Bernstein would have conducted the desert's recitatif with hunched shoulders and small movements of his forefingers pressing against his thumbs. *O friends, let's strike up something more pleasing. Joy*, the tenors sang. And the doves answered, *Joy. Joy.*

The astonishment of the "Ode to Joy" is that Beethoven was almost completely deaf when he wrote it. The cruelty of that fate is ironic and almost unbearable: "Oh Providence, give me one day of pure joy," he wrote as his hearing failed. But, unable to hear the joy in the nature he revered, he had to compose it himself.

I don't know how he did it. I've read that Beethoven could hear all the parts of a composition in the concert hall of his mind, rehearsing a symphony as he silently walked miles through the woods. And surely he could feel music in his body. All those years as an organist, wouldn't he have been shaken, body and soul, by the ecstasy of mighty organ pipes built expressly to shake Heaven and Earth?

When the Ninth Symphony was completed, Beethoven took it to Vienna for the premiere. He himself would conduct. Alarmed and not at all confident how that would go, the kapellmeister positioned Beethoven in front of the conductor's stand but gave the baton to the concertmaster, having instructed the orchestra and choirs to ignore the great composer as he waved his arms and flapped through the pages of his score.

It is said that Beethoven conducted with all the vigor of God conducting the angel choir—*Heaven and Earth will tremble*—stretching to his full height, dipping to his knees, flailing his arms, tossing his tousled hair. When the concertmaster pinched off the very last chord, the audience leapt to their feet, cheering and clapping, weeping, throwing their hats and handkerchiefs into the air. Beethoven conducted on through the tumult. Finally, the alto soloist put her arms around Beethoven's shoulders and turned him to the crowd, so he could see that the symphony had ended. How his body must have trembled in the uproar and acclaim.

AFTER THE SIZZLE OF A CAMPFIRE SUPPER, THE FRIED ONIONS and potatoes, after the balloon of the moon had floated above the dark desert mountains, Frank and I drove out to the sand dunes to listen for sidewinder rattlesnakes. The rising moon poured light on the sand, sharpening the shadow of every bush, each dent in the sand and lizard scrape. We moved carefully, even though the sidewinder is a small snake, only slightly longer than a piccolo, with far weaker venom than its nasty cousins.

It didn't take long to find a sidewinder's track. It can't be mistaken for anything else. The sidewinder moves by bracing her tail and the lower part of her body against the sand and throwing her head and body forward. The snake track is a series of diagonal lines drawn across the direction of the snake's movement.

We followed the track, expecting that where the marks disappeared, we would find the snake buried in sand with just its viper eyes sticking out, waiting in ambush for a careless lizard. Its jaw would rest on the ground to pick up tiny vibration waves speeding through sand from the padding paws of any passing mice. But no: We found the snake in debris under a stunted mesquite, perfectly patterned like debris under a stunted mesquite. In the disguise and shadow, we could barely make out its curve.

Our eyes couldn't help us much, so we listened. We could hear the rustle of snake scales across dry leaves, signaling that the snake was moving away. Too bad: Maybe we wouldn't get to hear the snake rattle this night. Then one of us—neither would confess—stomped a foot to throw vibrations at the snake.

It is impossible not to jump away from the sound of a rattlesnake. Not loud, but inherently alarming. Not a rattle, but a buzz. Not like a snake, but like an electrical short. We jumped back and crashed and ended up hugging each other to stay upright. That's what you do when you are surprised and unbalanced, don't you? You hold on tight to any available shoulders, jiggling with laughter. But *hist*. We froze, knowing that only our ears could tell us which way the snake was moving.

It was a beautiful night. The breeze smelled of dust and the strong, smoky scent of creosote bushes. Moonlight gleamed over the swells of the dunes. The air was cool on our sunburnt skin. In that absolute silence, we could hear the sand hiss under the slight weight of the snake throwing itself away from us across the dune.

IN HIS FAMOUS ESSAY "ON NATURE," RALPH WALDO EMERSON noted that few people can see nature, having only a superficial seeing. But for the dedicated watcher, "every hour and season yields its tribute of delight."

Crossing a bare common ... at twilight, under a clouded sky, without having in my thoughts any occurrence of special good fortune, I have enjoyed
a perfect exhilaration. I am glad to the brink of fear.... Standing on the
bare ground—my head bathed by the blithe air and uplifted into infinite
space—all mean egotism vanishes. I become a transparent eyeball; I am
nothing; I see all ...

I think Emerson means that sometimes, in joyful and unbidden moments, he loses his awareness of seeing, and simply sees. He loses his identity
as a seeing being and becomes sight. He doesn't try to see, or remark on his
seeing, or rehearse the sights. Seeing is all he is. A transparent eyeball.

If Emerson could become an eye, surely Beethoven could become an
ear. And could I also learn to listen so closely that I become nothing, losing
everything of myself except for the joy of sound? Could I stand on the bare
ground—my head bathed by the shivering air, my skull a bone-china bell,
glad to the brink of fear?

WE DROVE BACK TO CAMP FROM THE SAND DUNES, AND WHAT AN
awful violence that did to the silent night, a gasoline engine grinding out its
successive explosions, and wheels throwing gravel. The finale of the Ninth
Symphony begins with the "terror fanfare," when Beethoven throws his
D-minor chords on top of the B-flat chords and does terrible violence to the
meter. Trumpets blaze, drums snap, meteors most likely strike the Earth.
And then? Then quietly, the cellos and basses begin to play the perfect, simple, joyful theme, "as if they were teaching it to the orchestra," a reviewer
wrote, and gradually the choir and orchestra join in overflowing joy.

So it was, when we finally got to camp, turned off the engine, and
walked into the quiet night. We had arranged our sleeping bags on a sandstone slab over a stream that dripped into a round water pocket. When
we'd set up camp in the bright afternoon, the pool had been green with al-

gae, pocked with dragonflies, sizzling with the buzz of their wings. Canyon wrens had sung from a stony amphitheater across the stream, clear tones dripping down the musical scale in drops as liquid as the stream itself. But when we returned after our rattlesnake search, the night was utterly still and the pool was black as an eye, the reflection of the moon its shining pupil.

Frank climbed into his sleeping bag and promptly fell asleep. I sat on the overhanging ledge with my legs dangling off the edge. The wind had died, as it usually does in the desert night, and the cactuses were quiet. Coyotes were quiet too, deep into their hunt. The only sound I heard came now and then from a great horned owl some distance up the canyon. I could imagine the eager yellow moons of his eyes.

Every cactus stood in a black pool of moon-shadow close around its feet as the moon gleamed directly overhead. It was the supermoon, as full and as close to Earth as it would come all year. I watched the moon for what seemed like a long time. As far as I could tell, it didn't budge from the apex of the sky, but the black shadows under the cactuses filled and slowly overflowed east into the gravel.

I stretched out on the sandstone and closed my eyes. I could hear nothing but the pulse in my ears.

From deep in the canyon, the owl's exhalation echoed against rock.

A poorwill called. *Poorwillip.*

The moon hauled slowly through the night. As the Earth swung heavily on its bell-yoke, the night began to toll. Deep, soft belling seemed to roll over the plain, and ancient stone and ancient bone resounded. The saguaros sang out. Sand sifted down the flank of the moon-borne ridge. A drop of water popped onto the pool. Another. The poorwill called again.

Profondo, adagio, the music pulsed through the dark amphitheater. If I had been given the score, I would have gladly sung the alto part. But the music was unlike anything I had ever heard, and who wrote this symphony? I lay still, shaken by the night's extravagant expressions of profound joy. I would make myself silent and resonant, tuned to the wild Earth. I would put my arm around this composer and present him to the grateful world. I would leap to my feet and fling my hat into the air.

The International Union for the Conservation of Nature lists the sidewinder rattlesnake as a species of "least concern," with populations stable at about 100,000. The saguaro cactus, on the other hand, is hit hard by drought and spikes of extreme heat attributed to climate change. In Saguaro National Park, "establishment of young saguaros has nearly ceased since the early 1990s."

(International Union for the Conservation of Nature,
National Park Service)

Earth's Wild Music

DID DINOSAURS SING? DID THOSE SWAMPY, FERNY FORESTS RING
out with choruses of terrible, barrel-chested tenors? Did the velociraptor
hunt the steamy sky, screaming? Did the little mammals warble warnings,
as they scurried into cypress roots? Did the brontosaurus mother murmur to her new-hatched eggs? What a teeming, singing wilderness that
must have been, all those species thumping around, tuning up for the
next millennia.

Of course, dinosaurs sang, I thought. They are the ancestors of the singing birds and cousins to the roaring crocodiles. They are the progenitors of
the music the animals make; they are the Greek lyres of the animal king-
dom. Of course, dinosaurs sang. What clinched it for me was the feathers. All along, I had envisioned dinosaurs as gray-scaled or thick-skinned
creatures, naked and wrinkled as elephants. But new microscopic exam-
ination of fossils shows that they were splendid in colorful feathered capes
and plumed headdresses. The first time I saw a drawing of a *Tyrannosaurus
rex* with a thatch of gray feathers on his head and down his back, my eyes
were as astonished as his. Peacock blue-tailed, red-cockaded, green-masked,

striped, and spotted; dinosaurs were magnificent. If they had evolved colors to defend territories and attract mates or—who knows why?—to show off, then of course they would have sung.

Turns out, no. Turns out the syrinx, the organ that produces birdsong, and the larynx, the organ that produces operatic arias, didn't evolve until after the dinosaur extinction event. Turns out dinosaurs opened their jaws to kill but not so much to sing.

That's a letdown. But that doesn't mean they didn't make any noise. Maybe the forests didn't ring out with warbles, but they were raucous with animal sounds. If dinosaurs couldn't make music with their throats, they could still shake the heck out of the air. Some dinosaurs blew air into their closed mouths and through nasal cavities into resonance chambers, which we see in fossils as bony crests. They made the forest echo with clear, ominous tones, eerily like a cello. How do I know this? I have heard it, in recordings scientists made of the sound they produced when they blew air through crests constructed to mimic *lambeosaurus*'s.

Some dinosaurs hissed through closed teeth. The air rang with not only the heavy thud or quick scramble of clawed feet but also grunts and groans, chirps and rattles, or other "closed-mouth vocalizations," as scientists call them, emitted through the bulging skin of the neck. The *diplodocus* whipped its tail at supersonic speeds, cracking the air. I was just getting used to all this when I learned that some dinosaurs cooed to their mates. Like doves.

And now, get this. Turns out that even if dinosaurs didn't sing, they danced. "We know [that members of the family of *Tyrannosaurus rex*] had feathers and crests and good vision," paleontologist Martin Lockley said. "They were visual animals, but there's never been any actual physical evidence that their anatomy and behavior was co-opted for fairly energetic display. This is physical evidence."

The evidence he refers to is vigorous scrape marks found in groupings in 100-million-year-old Colorado sandstone. From the courting behavior of ostriches and grouse, scientists envision the dinosaur males coming together on *leks*, or courting grounds, bobbing and scratching, flaring their brilliant feathers and cooing. Imagine: huge animals, each weighing more than a dozen football teams, shaking the Earth for a chance at love.

I can imagine this—in a lesser kind of way—because, hidden behind a log wall, I have watched a football-sized bird boom and dance on its lek. It was the greater sage-grouse over in the Oregon desert, on a late-winter, snow-scoured, pre-dawn expedition when I thought I could actually freeze to death. Dinosaurs with or without feathers would have expired on the spot. But the birds were spectacular. Grubby little specked brown birds most of the time, on the lek, they bloomed. Their tail feathers spread into a pointy-tipped fan, revealing white-spotted feathers on their rumps. They draped their wings down over their legs. They shook the feathers around their necks into a fluffy white boa. Then, the boa parted as their chests swelled, exposing two big sacs like yellow breasts. Unlike any breasts I have ever seen, these sacs expanded and collapsed, giving out a loud boom. Sort of like *boomp, boomp, ba boomp ba.* Attracted from miles around, dowdy females wandered over the lek, acting unimpressed, like the pig judges at the 4H county fair. The males scratched the ground bare. Scale this up to dinosaur dimensions, and you sort of have what scientists envision.

What the story of the dinosaurs tells me is that if the Earth didn't have music, it would waste no time inventing it. And it did, with eager abandon and infinite creativity.

In birds, tantalizing evidence of birdsong is found in 67-million-year-old fossils, marking the first known appearance of the syrinx. The syrinx is the organ deep in birds' chests that they use to create their melodic songs. Now the whole Earth chimes, from deep in the sea to high in the atmosphere, with the sounds of snapping shrimp, singing mice, roaring whales, moaning bears, clattering dragonflies, and a fish calling like a foghorn. Who could catalog the astonishing oeuvre of the Earth? And more songs are being created every year.

The humpback whales in a given region sing a simple song that grows more complex as time passes. But at some point, they abandon that song and start a new melody, teaching it to others as they migrate. American robins learn their songs by imitating their elders, but after they leave the nest, they invent and improvise as freely as Ella Fitzgerald. The new songs of the magpies in Australia weave in the sirens of fire trucks, as wildfires rage. The whole planet sings.

Now, in the Cenozoic spring, a song sparrow raises its beak and throws its song to the sky. *Dee-dee-toodle oodle oo*. Human ears are built to hear its music. We hear most acutely in the range of 2.5 megahertz, which is the peak of birdsong. Human speech is pitched much lower, one kilohertz or below, and so is less central to our hearing. Why is this? Acoustic ecologist Gordon Hempton surmises that our bodies evolved not for cocktail party conversation but rather to harvest sounds from wild creatures. These are the aural signals on which our species' success depended: Birds chatting, unconcerned. Herds gathering. Corvids flocking. Sudden silence. They spoke clearly: *Here is safety. Here is water. Here is food. Here is danger.*

But that's just the beginning of the meaning we harvest by listening. Victor Hugo reminded us that "music expresses that which cannot be said and on which it is impossible to be silent." Listen. Breathe Earth's wild music into your body. You are not alone. Here is the harmony of which you are a part. Your joy is the exhalation of birds. Your fear is the pounding of hooves. The depth of your feeling is the depth of time. Your longing is a spring chorus of frogs, "the wordless voice of longing that resonates within us," as Robin Kimmerer wrote, "the longing to continue, to participate in the sacred life of the world."

Populations of the greater sage-grouse have fallen to 2 percent of their historic numbers, primarily because their habitats have been severely degraded or destroyed by fossil fuel development and other development on public and private land. Moreover, overgrazing and range mismanagement have reduced the number of seed-bearing plants on which the birds depend for food. Despite this, the sage-grouse is not an ESA-listed species.

(American Bird Conservancy)

The Sound of Human Longing

RAIN DRUMMED ON THE HATCHES AND SPLASHED OFF THE decks, but still we could make out the sound of a wolf howling from the cliffs over the cove where we dropped anchor. There was only one wolf; we listened carefully to be sure. The howl started low, leapt up, slid along the water, and sank away. Nothing answered the wolf's call. Frank and I listened, as the wolf must have listened, the question probing the clouds and damping out in the forest, the draperies of lichens and drooping hemlock boughs.

But the only response was rain pounding, then rivering down my sleeves and soaking my gloves. I tucked my hands into my sleeves, ducked my head, and hunched my shoulders to direct the water down my raincoat instead, to the deck of the boat and off the stern to the sea. The wolf howled again. I knelt to raise the anchor so we could drift closer to the cliff.

I knew the song the wolf sang. The first two tones made an augmented fourth, a dissonant interval, like the first two notes of "Maria" in *West Side Story*. It's an interval of yearning, of hope—the sound of human longing.

When my colleague, a concert pianist, explained the augmented fourth, she brought both hands together in front of her body, palms skyward, fingers spread, and lifted the air. For her, words are not enough to describe this interval. This is a sound that floods the soul, she said, and she strained forward from the waist. The augmented fourth is a heartbreaking interval, dissonance that comes so close to consonance, pulls itself so close, but never reaches the perfect fifth that is almost within its grasp.

She leaned over the keyboard and played two notes: C, F-sharp. Then she flooded the room with music made of the unfinished intervals, harmonies that lead toward resolution but never reach a place of peace. Tony, reaching for Maria. A Greek chorus pleading with the gods to have mercy on Orestes's soul, this man who has murdered his mother. Tristan, yearning for the white sail that will bring his beloved Isolde on a following wind. And Robert Schumann, poor lovesick Schumann, yearning for Clara. *Yearning*: this ancient word, diving straight through history from the beginnings of language itself, a word as old as *home* or *earth*. No one in Christian medieval Europe sang the augmented fourth, my colleague said. It was the *diabolus in musica*, the devil's chord, so powerful it could grab a parishioner, drag him to his knees, and pull him, scraping on the paving stones, straight to hell. And there I was in that tide-dragged island wilderness, also on my knees, trying to understand the pull of these same two notes.

I sat on my heels and strained to hear the wolf again, but the rain defeated me. There must have been three rainstorms stacked above us: A grayness in the air that wetted every surface, even under the canopy, soaking our hair but barely dimpling the water. An overloaded cloud dropping rain like sand from a shovel. And one unbearably heavy cloud that held the rain until it broke loose in huge drops that raised welts on the sea.

Listening intently, we pulled in our rockfish jigs and let the boat drift among small islands, until finally the dusk turned into dark. Then Frank started up the engine and slowly steered us back to the island where we had made camp.

THERE IS NO DARKER NIGHT THAN A NIGHT OF RAIN ON AN IS-land. Frank played his flashlight beam over the inlet to make sure the boat was still resting at anchor. I sat on an overturned bucket under a tarp stretched between hemlocks. Under my boots, the ground was springy, a thick layer of moss on a century of hemlock needles. Rain poured on the tarp, pooling in a corner that sagged until the edge of the tarp let loose, dashing the water to the ground. The tarp rebounded, spattering drops that sizzled against the lantern and wet my cheeks. I pulled my bucket closer to the center of the tarp. Even under its shelter, it was hard to stay out of the rain. Water bounced off the stems of highbush blueberries and salal, dripped from every stray end of rope, runneled the length of hemlock roots. I sat hunched, forearms resting on knees, and drank whiskey, closely rationed.

Somewhere people were laughing in brightly lit places that smelled of books and coffee. Families were sitting down to dinner, somewhere, and fishermen were making fast their boats in harbors, calling out to friends as they hoisted their gear bags to their shoulders and turned toward home. But there were no other people here and not another point of light for fifty miles in all directions. Tonight, just our little family, and in my flashlight beam, a narrow strip of island rapidly sinking into a flooding tide.

A loud mournful wail. I was on my feet, reaching for binoculars, but of course there was nothing to see in that darkness. It sounded again—a musical arch of three tones. I ducked past the tarp and groped to the edge of the island, and there was the call again. I recognized it as the wail of a common loon. Waking at night, the loon might have found itself suddenly alone, or have lost sight of its mate in the storm. It called again with frantic urgency; first, two sustained tones, the second higher and longer—two wavering tones on that rainy night after so many days of rain. Then it added another interval, even higher and longer. That was the wild, heartbreaking sound of the augmented fourth.

I yanked off my hood and turned my face toward the call. The loon flew toward me, then veered suddenly, and the cry slowly faded away. I strained forward, trying so hard to hear an answering call. What I heard was water on water and the slosh of tide on rock.

I should have felt a loneliness close to despair, there, in the night, in the rain, a thousand miles from home. What I felt instead was uncommon joy. What was there to long for, where all I wanted was what I suddenly had?— to be fully part of the night, joined by a song, by a simple shared song, to the loon, to the wolf, to the keening of all humankind, all of us together in this one infinite night, all of us floating in the same darkness, each of us, as we howl our loneliness, finding that we are not alone after all.

Lead pollution of the lakes where loons nest, drowning of chicks by pleasure craft, decreases in habitat, and the climate change–assisted increase in black-fly pests endanger common loons. Now loons are designated as either "threatened" or a "species of special concern" in many areas, "including much of New England (except Maine), the upper Great Lakes region (except Minnesota), and the western United States."

(Wildlife Land Trust)

Praise Songs I–VII

WHEN DARKNESS TURNS
UNEXPECTEDLY TO LIGHT

I. Glowworms

LATE ONE EVENING IN EARLY SUMMER, FRANK AND I SAT IN gathering dampness and dark, having just doused the fire where we had cooked dinner. We were on the western side of our meadow, near the oaks, looking into the swale where night collected first. Right there, at that bend in the landscape where the flat darkness of the meadow angled into the vertical darkness of the trees, I began to see clusters of tiny lights. They were unmistakably lights, although I doubted myself at first. They looked like constellations of stars in some sky I had never known, but why wouldn't there be strange skies that would reveal themselves at certain cosmic moments, pulling back the curtain of the forest duff?

I whispered to Frank, "Do you see that? By that alder?"

"I do."

We crept toward the swale, but as it turned out, we didn't need to have worried that we would set these stars to flight. The stars were the tail-ends of small grub-like things with black beetle backs and fleshy abdomens that glowed with cold light. We stood in the dark at the edge of the field, with dirt and soft-bodied stars in our hands.

There have been times in my life—I'll just say this straight out—there have been many times in my life when darkness turned unexpectedly to light. I would sing praise songs to these small eruptions of light. Praise songs, which are hymns, *hymnos*, which are psalms, *psallein*, which means plucked. Thousands of years ago, the plucking of the harp accompanied songs of praise. Pictures of the harps survive, as do many of these psalms, with cantillation signs indicating how to sing the songs, or with musical directions in superscripts: *With stringed instruments*, or *A song of love according to the melody of lilies.*

I would sing a song of praise for . . .

"Douglas-fir glowworms," Frank said that night. "I have heard of these. Luciferin causes the glow. Luciferin and luciferase, the enzyme that acts on it," and there was no reason to doubt him. Even when I asked why, the answer was forthcoming. "To attract males," Frank said. And why are males attracted to light? Because that's where they find the females.

Now, I love Frank, but this is not an answer. Maybe he didn't understand my question, which (as I think about it now) must have been, why is it so beautiful? There is necessary beauty in the world, I understand this. Beauty to attract mates, to attract prey, to attract pollinators. But so much of beauty seems to be bycatch, "unnecessary beauty," waste products of essential processes. The opalescence of the inside of an oyster shell, a rainbow around the moon, a baby's dreaming smile. Profligate beauty is a mystery to me. Sing praises.

II. Beavers

WHAT BECOMES OBVIOUS, PADDLING A CANOE IN THE MIDDLE OF the night, is how many different kinds of darkness there are in the world.

There's the furry darkness of spruce trees that mass at the side of the marsh. There's the slick black of the water itself, shining and smooth as the hood of a Buick. The night sky is dark like cracked leather. In darknesses like these, a paddler's back is going to be tense, ready to respond if the bow skids off a hidden log or plows into reeds. Of course I told myself not to flinch. Nothing will dump the canoe, I told myself, except your own fear, flinching. I pulled the paddle slowly along the gunwale, lifted it, pulled it again. Black water moved through the shadow of the boat. Then, right here, at the bow, a tremendous thud, like a rock heaved into the marsh. The water lifted.

It was a crown of white flames, I told Frank later, but cold. Created from dark water, it was a circlet of cold fire that could fit on a queen's head, something you would see on the cover of a cheap fantasy novel, but without the sword. "Darkness doesn't really turn to light," Frank said then. "It's a dark surface reflecting light from another source. Darkness doesn't set out to shine."

But maybe it does. How else can we understand why there is anything at all, except to say that at one astonishing moment, an infinitesimally small point of utter darkness set out to shine. And that is the only reason why Victor Hugo could truly write, "There is no such thing as nothingness, and zero does not exist. Everything is something. Nothing is nothing. Man lives more by affirmation than by bread." Sing praises.

III. Parkas

WINTER. TEN DEGREES BELOW ZERO IN OUR CAMP IN THE ROCK-ies. Clear, starry night. Frank and I had climbed into our sleeping bags early, wearing every article of clothing we brought. Silk long underwear, fleece shirt, down vest, wool muffler, down parka. Gradually, though, my sleeping bag warmed up until I was sweaty hot in all that stuff. Groping for the zipper, I wormed out of my down jacket, trying not to let in cold air. When I dragged the jacket out of the sleeping bag, it sparked and snapped, shooting stars into the dark. Excited, I pulled the jacket back in again. More crackling sparks. I stripped my muffler out of the sleeping bag. A river of sparks.

"Hold still, will you?" Frank said. "You're letting in the cold." But I pulled off the jacket, pulled it on again, struggling, wiggling, rubbing silk on nylon, until the whole dark tent shot sparks. It would be reasonable to think that the tent smelled like fireworks, but it smelled of ice on the needles of alpine firs. Sing praises.

IV. Spiders

IF YOU WANT TO KNOW HOW DARK A TRAIL CAN BE, TRY WALKING a jungle trail at night. The group of us had followed a flashlight to the edge of the river. With the lights doused, we crouched like crows, jerking our heads toward each chirp and squeal and sudden shout from the shadows. But it was too dark, much too dark to see anything. I don't know how much time went by. My eyes ached, trying to see—do eyes ever bleed with this trying?—and the muscles in my back crawled. "Can someone turn on a flashlight?" Frank flipped a switch. The beam found a tangle of lianas. The vines were full of eyes, pinprick silver eyes, all of them looking at us. Eyes, like a thousand tiny coyotes. Paired eyes, all staring. "They're spiders," Frank said. "We're seeing the eye-shine of spiders." The spiders stared at us, and we stared back. Can't spiders blink, for god's sake?

If spiders were to turn a flashlight on us, we would not return their silvery stare. Human eyes at night are blank dark pools, taking it all in and giving nothing back. Fish eyes shine white, horse eyes shine blue, raccoons' green, coyotes' red. But humans'? Flat-black, felted, obscure: like the dark matter of the universe, soaking up the light of the stars. Sing praises.

V. Lightning

ON A NIGHT TINGED AT THE EDGE WITH RED, WE PUSHED THE canoe through tule reeds on a mountain lake. Our movement startled a red-winged blackbird that chirped and rustled the reeds. Pushing past a stick, the canoe squeaked. Then the lake went dead quiet. There had been fires in these lodgepole forests, and the full moon was as rusty as a wreck. From the hills around the marsh, we heard rumbles from distant lightning, and we stopped our paddles to listen, rocking in the ruddy dark. Then, suddenly, the darkness broke open in a jagged white crack that ripped from the hills to the crest of the sky. Astonishment, a sharp blow to the mind, lit everything with unexpected light, and all the world suddenly came clear. A blast of thunder rolled over the lake, engulfed us, and roared away to the east. Then there was nothing but that dark red silence—and a spot of fire burning like a candle-flame in the top of a pine on a ridge across the lake. It began to rain. Sing praises.

VI. Bluegills

SO THERE IS THIS. MANY DECADES AGO, I WAS FLOATING ON MY back in an Ohio quarry. The sky was purple, the water was warm, and the moon was milky white. I don't know what lie I had told my parents, that I was allowed to swim here in the dark. I do remember thinking I could sleep here, floating, and I tried to make myself as still and insubstantial as a dream. But a bluegill rose under me and nipped my leg, and I changed my mind and floated upright, with my head out of the water mostly, and my legs pumping to turn me round and round, so I could see the dark weed bed, then the beams of headlights through the maples, then the black forest. What happened next, I couldn't explain at first. A point of light popped on the glossy black water, and concentric rings spread from that point, expanding and glossy. Then another point of light rose and sent out its rings, and another—all around me, expanding rings of light. The rings swirled and coalesced, then spun away, twisting off new galaxies of light.

Thinking about it now, I imagine it was bluegills, rising to sip insects, disturbing water that caught the light of the moon. But back then, I thought, this looks like simulations I have seen of how the universe began, these expanding rings of light. I am floating in the middle of the Big Bang, which is quieter and gentler than I would have thought. Sing praises.

VII. *Jellyfish*

NIGHT HAD FALLEN FASTER THAN WE EXPECTED ON THIS DESERT island, so we were forced to find our way back to camp by following the water along the beach. This was not as easy as it might seem. I stumbled, and before I caught myself, I had splashed into the sea. Damn, because sneakers soaked in saltwater will never dry. But look. In the disturbed water, small lights bobbed along like luminous coracles.

I swept my foot through the water. All around the floating shoelaces bobbed a flotilla of white lights. "Jellyfish," Frank said, bioluminescent jellyfish. The lights throbbed by, pulsing as slowly as a sleeping heart. Do these jellyfish have minds, or are they the mind of the ocean? I must have said this aloud, because Frank turned to me. "Minds? Jellyfish don't even have brains," he said.

They have a neural network, a hairnet of nerves in the shine of the jelly. Something touches the jellyfish—maybe a surge of water, or a shoelace— and a nerve fires, then others, all around. The nerves flip from a high level of energy to a lower level, and the excess energy flashes out as light. So. It's awareness of a touch that lights up the jellyfish, I remember thinking. We are watching the awareness of a jellyfish pulse and glow in the saltwater of a shallow bay.

Never mind about camp. I pulled Frank down onto the beach, lay my head on his shoulder. Unseen jellyfish flowed through dark shallows. The Milky Way flowed through galaxies. And flowing between us? I like to think it was the same infinite stream of awareness, throwing sparks. Sing praises.

In a ten-year study in Germany, researchers found that both the biomass and diversity of species of spiders and insects declined by about one-third in all regions. Others note that habitat destruction, such as forest clear-cutting and pesticide use, reduce spider numbers and change their activity.

(Technical University of Munich)

The Love Child
of Father Time
and Mother Earth

YESTERDAY WAS THE LONGEST DAY OF THE YEAR. AT OUR ALASKA cabin, the sun shone for eighteen hours and sixteen minutes, sunrise to sunset. The steady increase in light building up to that splendid achievement, day after day, made me just plain happy. I was even happy when a red-breasted sapsucker pounded on our metal roof at 2:30 in the morning. That woke me enough to see watery pre-dawn light seep between the hemlock boughs and to hear what came after the sapsucker. All three thrushes sang—hermit, Swainson's, and varied—and the Pacific wrens trilled along. Sparrows sang—not the sweet, jumbled whistles of adult birds, but the experiments of the yearlings, who did fair to poor imitations of their elders with an exuberance and lack of embarrassment that made me smile in bed.

Smiling in bed, I think, is a very good thing. In time, the little birds would get it right.

The sapsucker's *rat-a-tat* woke me again at 3:51 a.m., one minute after sunrise. The drumbeat was sharp and insistent. What good luck it was for him, to find the drumhead of our metal roof to amplify sound far out of proportion to his scrawny little bird body. It must be powerfully rewarding for something so small to make a sound so large. It hadn't occurred to me that our cabin was designed in the long traditions of drums, but of course it was. From now on, I will call the cabin the Drum House.

When I told my sister about our early morning drum concert, she sent a frowny face and wrote, "You could get a white-noise machine or wear earplugs to keep from waking up." But we already have a little creek on each side of us, and after a rain, we sleep to sweet swash and splash. And never would I trade the morning chorus for an extra few hours of sleep and my stupid little dreams about losing the keys to my filing cabinet or hiding coins in the snowy berm of a forest road. Also, I told her, I love the slip-and-drowse consciousness of being half awake. She emailed me a thumbs-up.

Beside me in bed, listening to the sapsucker, Frank said, "It's amazing that the little guy doesn't give himself a concussion." And it's true that when I've watched him—gripping the sill with gangly claws, propping himself on his tail, hunching over the roof, banging away with his beak as if he were typing some manifesto of birdness that was important enough to save the world—I have wondered.

"Its beak is uneven," Frank said. Understand that Frank sleeps with the sleeve of a black sweatshirt across his eyes to keep out the light, so the voice discoursing on sapsuckers came from a big biologist lying on his back, looking like the Lone Ranger. "The top bill is longer and not so strongly connected to the skull. The lower one is heavier. It sends the vibrations down away from his brain."

And then there's the thick, spongy skull, and the brain tightly packed into it, so it doesn't slosh around in a sea of cerebrospinal fluid and smack on the skull, back and forth, like a loose skiff between the dock and the break-wall in a gale.

"Also," Frank said. I rolled over to face him. "Why do you think the sapsucker taps that rhythm?" He pulled the sleeve off his face. "*Brumdiddy-diddydiddy. Deet. Deet.*" He paused for a long time and then did it again.

Four o'clock in the morning, and this man was lying with his eyes closed, doing an adequate imitation of a woodpecker.

"I don't know. I give up." I knew the answer was going to be wilder than I am capable of imagining.

"So the pounding creates heat energy. Right? Like nails get hot when they are pounded. It's a bad fever when your brain heats up. So the sapsucker has to count time between riffs to let his brain cool down."

Before this conversation, I had been thinking maybe I would go back to sleep, but at this point, there was no chance. Who thinks of these things? But Frank had one more.

Turns out that the little sapsucker has a special hyoid bone like a strap that comes up along the back of his skull, wraps through his nostrils, and connects again to the back of the skull, a kind of seatbelt that keeps the skull from whiplashing and transfers the vibrations away from the beak to the body. Now that is impressive, a very neat trick.

Frank plummeted back to sleep, but I lay sort of dazed and dozing, dreaming of Father Time and Mother Nature biting and kissing as they rolled down the beach—the fierce battle between the forces of destruction and the forces of creation. Father Time had a tangled beard and dirty white robe, but somehow he had lost his shepherd's crook, and she wore nothing at all, a hefty, ripe old woman.

Oh, they love each other and they hate each other and they need each other desperately. Father Time, the agent of crumble and disruption, endless depletion, mass dying, selective dying, a force hungry enough to eat mountains. Mother Nature, inevitable creation and imaginative flowering, endless giving. She tries that, or better, this—another unfolding, maybe a twist in the tune or another color, maybe red this time. Through millions of years of no-holds-barred evolution, they wrestle and scold and make wild love. Together, they create new life forms, ongoing. And somehow, they figured out the red-breasted sapsucker, the Ringo Starr of birds. Who would have thought?

I have to pause a moment to admire evolution. You can't think these things up. Not even God would jerry-rig a hyoid bone. Not even God would think to tell a sapsucker that if he pecked holes in a hemlock's bark, the sap would ooze out, sticky enough to catch bugs and hold them there, so the sapsucker could swing by on the way home from work and pick up supper. And what engineer would think to tighten up the sapsucker's skull so he didn't brain himself making the holes? And could even God imagine a rufous hummingbird to follow the sapsucker around, sipping from his sap-wells? Even if you are omniscient, knowing everything, that doesn't help you when things don't yet exist to know about. But the urgency of life and the inevitability of death over 4.3 billion years made this bird.

The evolutionary process reminds me of backwoods Alaskans, endlessly brilliant in creating and fixing, using whatever limited stuff they have in the shed. Get an empty juice bottle and some wire and start tinkering. Try something. If it serves some purpose, keep tweaking until it works better. If it doesn't, throw it out. Tweak this, replace that, try to make this bolt tighten a joint it wasn't intended for, don't be embarrassed if you don't make it nice the first time. You have another couple of billion years to get it good. And at the same time, start in on the plumbing of hemlock sap and green-blooded ants, but without any idea what you are making or why. I look out the window of the Drum House into a frigging miracle of improbability.

ON THE DAY OF THE SUMMER SOLSTICE, WORKERS FLOATED IN on the rising tide with tool belts and climbing gear. They climbed onto the roof of the Drum House, tore off the old metal sheeting, and replaced it with nice gray high-tech metal roofing. Around the edge, they installed flashing to direct water from the roof down the pilings. What the sapsucker thought of the power drivers and hammers, I can't imagine. But I heard him pounding bravely on a dead spruce by the water, maybe terrified that his wife would fall for the new guy with a really big hammer, even though the bird had a white mustache nicer than the carpenter's.

The next morning, scheduled to come eighteen seconds later than the morning before, the sapsucker did not play the drums on our cabin and we slept until 8:00 a.m., waking cross and confused. I studied the roofline with binoculars. The new flashing made it impossible for the sapsucker to hold tight to the rafter, and he couldn't even tap.

Without thinking, we had traded twelve thousand dollars and a new roof for the sapsucker's morning drum tattoo.

Damn, the trades we make.

I've been frantically nailing scraps of two-by-fours and one-bys to the soffits to give the sapsuckers a grip. I salvaged some random pieces of roofing to screw to the siding. Frank does not approve of this, it's true, thinking that inviting a bird to peck your house is not the greatest idea. But it's a moot point. A few days have gone by and the sapsucker has not returned. Now the rains have moved back in, so the day imperceptibly dims toward night, which is only a deeper shade of gray than the morning. Even the rufous hummingbirds are gone from the feeders, and Frank is down under the cabin with an empty juice bottle, some wires, and odd pieces of PVC pipe, trying to direct water from the new downspouts away from our pilings and out toward the creek. It's not going well, and I wish he would take a break to let his brain steam off the excess heat.

You know what else? Even the rain is quieter on this new roof, Frank says, because the gutters no longer leak. Okay, but I miss that music. I guess if you're going to make a good trade, you should be very clear about what you are giving up and what you are getting. Maybe that's Rule One. When I trade a new roof for the music of a sapsucker, it's not just that sapsucker but the splendor of his evolution and the promise of evolution to come. It's a value beyond even value itself. And besides the harm to the bird, which is consequential, maybe the greater cost is to my integrity—saying I care about the bird, and then making an unthinking trade where the bird comes out the loser.

And that leads me to Rule Two. To make a good trade, you should be sure that what you are trading away is yours to trade. What have I taken from the sapsucker and his little hanger-on, the hummingbird? We had come to a cross-species musical collaboration, and now I have taken it away. What have I taken from my grandkids, who would have riffed on the drumbeat with little fists on the bunkbeds, scatting along until their parents convinced them that it wasn't morning yet, despite the light? The pattern of trades in a capitalist extraction economy looks more like stealing than trading. Yes, I want this (this money, this stuff, this oil or coal), so here, I will trade it for something that belongs to someone else (that old man's labor, that fresh water, that young joy, that child's health, that planetary thriving, that animal's nestlings, that marshland music, that run of salmon).

I'm sorry. It's just morning, and I'm already ranting.

PEOPLE SAY THERE ISN'T ANY OTHER WAY. THIS IS HOW IT IS: everyone seeking his own advantage in whatever trades present themselves— that's what makes the world go around. That is baloney. There are other ways, especially in ecosystems that haven't been ruined, that still offer an abundance, overwhelming.

Yesterday, my neighbor Tom walked a half mile from his cabin to ours to give us a bag of kale from his garden, braving the teenage bear that hangs out in the curve of the beach. The day before, Frank and I had run our skiff up to Tom's beach on our way home from fishing to drop off a halibut filet. Frank walked up to his cabin with the filet draped over his hand. When Nick came to work on the roof, he brought a bag overflowing with green onions and a dried-milk can filled with blue bachelor buttons. In each case, they had more than they needed and wanted to share. There is an alternative to stupid, selfish trades. It is sharing, what our mothers teach us before kindergarten, and why do we forget? Why do we abandon that model so completely that we think it's an impossible way to live? Why, when it's how people have successfully lived in this salmon haven for generations?

This afternoon, I am going to skiff in to the little harbor and walk out to the community garden. There are two signs on the fence. One says, "Free food for the doing. Mind the bears." The other says, "Be sure to close the gate," as if anyone needed to be reminded, with a young bear roaming the beach. It's a beautiful garden, neatly weeded, abundant with food. I will fill my bags with spinach and bok choy. Maybe I will cut some broccoli. Then I will spend a couple of hours digging weeds and tossing them over the fence. Next week, there will be peas and small yellow squashes, and more weeds to pull, and maybe it will be time to thin the carrots. Everyone is trying to give me rhubarb. I will be glad to bring good food back home to Frank, exactly what he needs.

Here's what I want to say. This is not an insufficient world, where every gain is a loss, and every loss is a gain. We are not born to self-dealing. The world gives freely, sharing the birdsong, the green nourishing food. Until human populations overwhelm their bioregions, there will be enough to go around, if no one takes more than their fair share. If no one tries to have it all. If no one wrecks the chances of replenishment.

DO YOU KNOW THAT THERE IS A WEBSITE CALLED "HOW TO GET Rid of Sapsuckers"? Can you believe that?

- Step one: Hang reflective items (CDs, tin pie plate, wind chimes) to scare off the sapsuckers.
- Step two: When you see a sapsucker, go outside and scream at it. If you have a toy cap gun, shoot it in the bird's direction.
- Step three: Spray sapsuckers with a garden hose when you see them.

- Step four: Cover trees with sticky bird deterrent, which will get on the sapsucker's feet and discourage it.

I told my husband that the website should be called "Four Easy Ways to be a Complete Dick."

Then I felt like a bad person, because we got rid of sapsuckers without even trying.

IT'S A WEEK AFTER THE SOLSTICE, AND NOW WE ARE LOSING A minute of sunlight each morning. Soon we will fall off the top of the curve that plots daylight, and we will plummet toward the dark days. So far, the sapsucker has not returned. Oh, he tried a couple of days ago, but he couldn't get a grip and the taps were pathetic. I don't know if there is still time for a second brood of sapsucker babies anyway, or if his mate will deem him worthy of the effort. Two rufous hummingbirds are at the feeder, but they just fight, driving each other away. Two feeders, sixteen perches, two quarts of sugar water—and they fight over one lousy beak hole. Nevertheless, the varied thrushes are getting to be prodigies of the tremolo whistle, and their beauty cheers me up.

Wait a minute. He's here. He's tried a few taps on the porch rafter, and now he's doing his best at the peak of the roof. Never mind. He's gone again. But I have a new idea for making him a sounding board. It involves some roofing screws the roofers lost between the boards of the deck and a piece of scrap metal. Wait, maybe there are two sapsuckers now. One is pounding away on the porch, loud and strong. And another one is doing a mismatched imitation up on the peak of the roof. I'm thinking this might be his son, learning the trade. And maybe his son has a sharper claw or stronger tail, and the son of his son will have a fancier mustache or more complicated rhythm, and maybe the sapsuckers will bring a new jazz improvisation to the woods, and the hummingbirds will buzz along, and the thrushes will scat. Then old Father Time and Mother Nature can turn their attention to the necessary further evolution of human beings, because maybe we too can figure out how to move to the rhythm of the sapsucker, given enough time and improvisation.

Numbers of red-breasted sapsuckers are declining slowly, as northwest forests are repeatedly cut. If average global temperature rises by 1.5°C, the sapsuckers will gain 18 percent of their range as the birds move north, and maintain 34 percent, but lose 66 percent as spring nesting seasons become too warm for the chicks.

(Audubon)

Repeat the Sounding Joy

> While fields and floods, rocks, hills, and plains
> Repeat the sounding joy,
> Repeat the sounding joy,
> Repeat, repeat the sounding joy.
> —"Joy to the World," verse 2,
> ISAAC WATTS, 1741

THE MUSIC AND WORDS OF "JOY TO THE WORLD" MAKE ME think that the carol was meant to be sung on a snow-glazed morning, as crimson and azure stained-glass windows pour light on golden curtains hung at the altar and bells ring glad tidings over white hills. But when I let the song lift me into my memories of the times when I have truly felt the joy of the world, the music takes me first to a tawny field lit by late afternoon sun.

Fields

POINT REYES, CALIFORNIA

I WAS FOLLOWING A WALKING TRAIL THROUGH A COW PASTURE, although this was not my preference. Cows intimidate me. I am leery of their pushiness. But they retreated until they were pressed against the fence, and I proceeded toward the ditch and the barbed wire on the other side of the field. Last year's grass, laid down in swirls by wind and rain, was tawny, like a cowlick on a kid's head. The mowed path was green as lawn. I saw something moving in the ditch. Dog? Coyote? Fox? I wished it would reveal more of itself. It did. The animal climbed into tall grass on the rim of the ditch. Now I could see the sway of its back and its muscled hips. The animal was longer than I would have expected, and the way the back articulated with the hips, loose-jointed like a lion, said cougar. I wished it would show its tail. It did—a long sweep of tail, tipped with black. Cougar: yes. It made a small leap to the top of the ditch to show me its full profile, yellow as the dried grass. Then it turned its head slowly and looked me dead in the eye.

There was most of a field between us. I don't know the distance. The cougar could probably have measured it in seven powerful leaps. Next to me was a stock tank. The corrugated tin wall was chest high, but I was sure I could climb it. Even though the water was tarry and green-glazed, that's where I would go if I needed to, hoping it was true what I had heard some place or other, that cats are afraid of water.

The cougar sat on its haunches, lowered its head between its shoulders, and watched me. I sat in the grass, crossed my legs, and watched it. It was hard for me to sit still, I was so excited. What I wanted to do was run, shouting, "There's a cougar in the cow field," so everyone could come and see. But I know not to run from a cougar, and besides, what I really wanted to do was stay in that place, to see if I could be more patient than a cougar. This turned out not to be the contest the cougar had in mind. The cougar slouched to a stand and started to saunter along the fence line, casually, as if it were smoking a cigarette. This took it across the line of my direction, taking it no farther from me, bringing it no closer. But I could see that the fence line was about to take a ninety-degree turn, and if the cougar followed that, it would be heading straight for me.

And it did, making the turn and slowly moving through tall grass, dropping its head now and then, or raising its muzzle. I could hear the cows crowding against the far fence, that clump and huff, and house sparrows chirping at the base of the water tank. I thought I heard the cougar's paws rasp through dry grass, but that could have been my breathing. Halfway across the field, it stopped again, sat again. And so we watched each other, until the sun dropped behind the oaks, shooting a shadow across the grass, and I thought, as I had been thinking all along, what is going to happen next?

Five o'clock frightens me when I'm away from home. At twilight, I want to know where I will be sleeping, and how I will get there. That's when I ask Frank to pull off the river and find a campsite. Or when I reach for my cellphone and call ahead for a hotel. When the sun goes down, everything changes. At work, my office window, which in daylight is a cornea that lets me see the world, becomes a cornea that lets the world see me. I felt that sudden switch in the field, when I looked again at the cougar and it was still looking at me. I stood up very slowly, graceless on feet gone stiff and fuzzy, never taking my eyes off the cougar. I backed up a step, another step, another, all the way back across the field, until the cougar was just a tawny spot, still sitting in the grass next to the fencerow. All the way, I thought how I would spread the news.

I told the man sitting on the curb in the parking lot, lacing his running shoes. I told the young couple drinking Cokes in their car. I told a woman getting out of the passenger seat to open a gate. I called Frank and left him a message. I wrote it in my journal. There is a cougar in the cow field.

Floods
ARTIC, WASHINGTON

TROOPERS GOT ONE LANE OF HIGHWAY 101 OPEN AGAIN ONLY A few days after the flood. Water still crossed the road in low places, and we had to dodge rockslides and shattered branches. It was raining again and overcast. Flashing yellow lights of highway crews and electrical repair trucks smeared in each swipe of our windshield wiper blades. Just past the Artic Tavern, we turned into the trailer park where Frank's brother lives. This is no paved parking lot. This is the wild Pacific edge, a few silver trailers tucked under immense, moss-darkened Douglas-fir, back in the huckleberries and ferns.

In the aftermath of the storm, the trailer park looked like a logging deck—black mud, felled trees and shattered limbs cut to stubs, piles of yellow sawdust and neat stacks of new cordwood. Mud still splattered the sides of trailers, thrown up by root wads torn from the dirt by the weight of falling trees. John was there to meet us—a tall, square-faced man with a silver braid down his back. When I rolled down the window to give him a hug, the sweet smells of evergreens and cigarette smoke flooded in. We hadn't seen John for a long time, and he looked good.

We walked together up to the tavern, hopping puddles and windrows of flood debris. The stories started even before the woman at the bar could draw the beers.

"We all holed up in the tavern," John said. "I mean, it was taking your life in your hands to go outside. All that rain loosened up the roots, and then when the gale blew in, those trees just went over, one tree taking down the next. We could hear them go. It'd be spooky quiet, then we could hear the wind starting to come, just this whisper, then, man, it was like a freight train was headed right for us, everything shaking and the candle-flames blowing straight to the side, and then *crack* a tree let loose, snapping its roots and crashing down. We could feel it through the floor, and we'd listen real close to see if the sound was cracking wood or splitting aluminum, and whose place was going next."

John scooted over to make room for another one of his friends. There must have been five of them at the table by now, gray-haired men, easy in their chairs.

"My buddy says, you gotta help me, man. I gotta get my truck out before it gets smashed. So I got my peavey and my chainsaw and we went out in that weather. Dark. Man! It was crazy—no wind, just this smell of cedar trees so thick you could drink it—and then we got slammed with wind so strong my buddy grabs me and we're running along with the wind, trying not to. Wind smacked us into the side of a pickup, couldn't help it. The river had crossed the highway and ran down the road into the trailer park, pushing all this stuff. Limbs and electric wires. Trees are down everywhere. My buddy has a flashlight, but we can't tell what's ground and what's flood, everything covered in blowdown—ferns and moss and limbs and aluminum siding. We'd hear a tree crack and we'd tuck our heads down into our necks, like that's going to keep a limb from pile-driving us."

Somebody in the bar laughed. Pool balls cracked. The fire clanked and fell onto itself. One of the men got up to throw in another log.

"There's no way, I told my buddy, but he says, just crank up that saw and clear the road. So I'm feeling around with my chainsaw, trying to limb up this tree—might as well have had my eyes closed, it's that dark, except for his headlights, just about blind me—and I'm rolling slices out of the road, clearing a section, and he gets his truck started up and jams out the road, rooster-tail like a Ski-Doo. Parks it right in the middle of the Highway 101. Figures that's the safest place.

"That night, we got up a pool tournament in the tavern, setting the candles right there on the pool table. People kept coming in. The Comcast guy, he figured maybe it's safer to drink beer with us than ride a cherry-picker in the storm. Annie, that's her over there. Monica and Mike never could get through to Hoquiam, so they slept on the tables by the woodstove. Wind blew the fire back down the stovepipe sometimes, spraying sparks. When people stopped coming in, we knew the North Fork'd cut us off."

John rested on his elbows, looking into his beer, grinning.

"I tell you, I never knew anything that shook me like that wind."

Rocks

KARTCHNER CAVERNS, ARIZONA

I DON'T GO INTO CAVES. THEY SCARE THE BEJEEZUS OUT OF ME. The earth moves. That is a fact. Openings slide shut. Rocks fall. "You'll have to go some distance to get me into a cave," I tell my daughter. And then she does. She announces that her entire wedding party is going to take a tour of the cave, and the family will be the guests of her new mother-in-law. This is too cruel.

Understand how nice it is on top of the hill—blue sky, cold and clear. Yellow rabbitbrush still glows, even with a brush of snow in its shadow. Barrel cactus are fat and shiny with rain. Cactus wrens are chirring. The wind smells of woodsmoke and sage. And then, right in the middle of this paradise, there's a thick metal door into the side of the mountain, like the door to a bank vault or the entrance to hell.

Okay, I say to myself. *You are able to go through a door.* Tour guides close it behind us, and here is a long tile-lined passageway with another door at the far end. *Okay, you've changed terminals at O'Hare; you can go through an underground tunnel.* There's the second door. *Okay. You can do it; a door is just a door.* They close that one behind me. I wheel around to make sure I know how to open the latch, even in the dark. *Okay.* But suddenly I can't breathe. The air is too wet. How wet can air be, before you drown? *Okay.* I suck in my breath. The lights go off behind us. *Okay. It's okay.*

Simultaneously, lights go on in front of us, and we are in a high-vaulted cavern made of stone that glistens like sugar. Pillars delicate as soda straws reach to a ceiling that seems to be sliding in sparkling shields down the walls and dripping light from each point and fault. The floor falls away in glowing moon-milk terraces, then rises suddenly to a throne studded with stars. A fin of rock glows blue as a glacier.

I had no idea. I had no idea there was an astonishing world inside the scratchy hill. I had imagined the cave was made of rock, not glass. The others wander on, and I'm glad for them to go, because I want to be alone in this vast, shining silence. I lean on the railing and wait. A drop of water falls into a pool. *Tock*. Another. *Pock*. Water rings against rock and echoes through the hall. Water chimes against stone and sings like glass. I didn't know there was such music in these hills. I didn't know I was crying. So much takes me by surprise.

Hills
BAYFIELD COUNTY, MINNESOTA

THE ROAD GOES GENTLY UP A HILL BETWEEN TWO ROWS OF white pines. Everything—the road, the hill, the pines—is blanketed in snow. The snow is blue this late at night, except where a flake catches some cosmic light and glistens white, and except for our tracks, which are white as well. The black sky seems almost too cold for stars. We have left the pickup at the gate, at least a mile behind us now, and trudged up here, three women who have only just met. We are going into the hills to sing for wolves.

I have howled for wolves before, often standing at some random edge of ice in the dark, alternately caterwauling and listening, trying to arouse a territorial response from the night-bound lake. But this is different. Locals already know that there is a family of wolves denning a few miles up the road. We don't need to find them. Instead, our job is to count the pups. So here we are in our boots and mittens and mufflers, hiking through the dark toward a family of wolves, trying to make no sound at all, but squeaking like mice on the cold snow. I wouldn't mind squeaking like a lumberjack or a hunter, but our sound is distinctly wee and succulent. If a wolf were to stalk us, we would never see it in the deep wells of darkness under the trees, and we would never hear it over the rhythmic creak of our boots.

"Maybe," I say, "we could hold hands," and we do, finding some comfort as our moist mittens freeze together.

We tromp along. Under the snow, the hill slopes smoothly away on both sides of the trail. Once, we startle an owl from its perch. We stop to watch it sail over its star-shadow. We walk more slowly as we get closer to where our leader thinks the wolves will be. At last, she signals us to stop. We stand quietly until even the memory of the sound of our movement fades away. Her mitten taps one, two, three, to remind us that we are to count small voices. Then she begins to sing.

She makes a soft sound, as if a mother were trying to put a baby to sleep. The song wanders up and down a minor scale, a tuneless lullaby. Then, right at our feet, scarcely a yard away, high-pitched cries—*one, two*, but I have no idea how many pups are yipping, and "oh my god, we're too close," and we are stumbling backward down the trail. We turn and run down the corridor of snowy pines, tripping on our footsteps in the snow.

When we stop, the voices of the pups have fallen away. The trees open in front of us. The black night domes over our heads. Stars glimmer in the snow. In that deep quiet, I can hear the sounding joy, belling across the snow-covered hills, and I don't know if it's the hills resounding or if it's my own heart that is ringing for joy.

Night wind shakes the stars in the trees, snow sings off the slope of the hill, wolves hum to their pups, and the depth of the universe throbs like a gong. Somewhere in this same night, choirs raise their voices in "Joy to the World," shivering the candle flames in the great cathedrals, and mothers sing the words softly to their children after they turn off the lights. *While fields and floods / Rocks, hills, and plains / Repeat the sounding joy.*

Repeat the sounding joy. The more hollow a heart, the more resonant it can become. I would make of this body, this life, a sounding board, tuned to that sympathetic vibration, which is sympathy, which is feeling together, which is compassion for all the world.

Plains

SASKATOON, SASKATCHEWAN

THE YEAR OUR DAUGHTER WAS BORN, WE RODE THE TRAIN FROM British Columbia to Ohio for Christmas. We had a tiny berth, with two bunk beds and a window the length of the room. It was dark when we set out from the city. I remember lying on my side with the baby cupped in the curve of my body. I watched the red and green and yellow lights of the city give way to moonlight on the cliffs of the Fraser River canyon. All the evergreens were weighted down with snow, and the river, far below us, was a silver thread through white drifts. On the switchbacks, the whistle of our own engine sounded sometimes in front of us, sometimes above us, sometimes behind, always far away. We climbed into the Rockies in the early morning, snow falling heavily, white through the white trunks of the aspens against the white sky. A single cross-country skier stood at a railroad crossing, wearing a red sweater.

Our baby was crying, pulling at her ears. We knew they hurt her, and I remember how worried we were, so far from home. I bundled her close and nursed her as the train descended onto the plains, and the dark descended, and the snow fell and fell. Then we were moving through the night across endless snow-blanketed plains, dark except for a white tunnel of light from the train's headlamp, visible when the train made a broad curve to the south. Then there was nothing for miles and miles but darkness and a crying child, and the rumble of the train. I sang Christmas carols, *and heaven and nature sing*, and rocked the baby in the rhythm of the thump and drop of the wheels. After a long time, she stopped crying and went to sleep.

For hours, I held her close and watched snow blow from the darkness against the window of the train. I must have slept. When I woke, I could see small globes of yellow light glowing through the snow, lamplight thrown from the windows of farmhouses evenly spaced across the plains, coming together and moving apart as the train approached and passed. At every crossroads, bells rang and the train sang out, and I could imagine that the people in the little houses felt themselves warm and glad to hear the music from the snow-muffled night.

Between 7,700 and 11,200 wolves live in Alaska. In 2019, 1,300 were shot or trapped by hunters or sportsmen. An additional two hundred were killed by Alaska Department of Fish and Game officials or their proxies, often from airplanes, bringing the total annual kill to an estimated 16 percent of the population.

(Alaska Department of Fish and Game)

Songs in the Night

WE HAD COME LATE TO THE COASTAL DUNES, BACKPACKING IN BY the unreliable light of headlamps. It's a challenge, pitching a tent in the dark, throwing sabers of light everywhere you look, knowing that if you put down a tent pole or a stake, it will be lost until moonrise or morning. But we get the tent up and then sit on our packs in front of the door, listening to the night.

Whenever Frank crawls into a tent, he falls asleep like a bag of sand dropping off a truck. I think the whole point of sleeping out is to stay awake as long as I can, listening. Sometimes I brag to Frank that he could lead me anywhere blindfolded, or with a pillowcase over my head, and I would know by the sounds what time it was, and where he had pitched the tent. He points out that he usually knows where he is, and it doesn't matter what time it is, if he is happily asleep. But I know he wakes up in the night and listens, and he wonders what makes the sounds—what animal is calling, and by what physical process the sound comes to be. He's a scientist, after all, a student of the chemical pathways through which an external stimulus enters the brain and stirs things up.

The wind sifts dry reeds in front of the tent, tiny claws scrabble against rough bark behind it. Tree frogs clatter from all directions. A killdeer calls. Surf rumbles like a thunderstorm on a far horizon, and close at hand, water slips softly in and out over sand, like the breath of a person who is not afraid. So I infer that it is late, maybe midnight, on a freshwater lake behind the foredune on the Oregon coast, on sand between shore pines and a marsh. "Good guess," Frank says, looking out over the marsh in the moonlight and the dunes beyond.

I'm not especially good at recognizing birds by their calls, but I do know the killdeer. It calls its name in the night, not exactly *killdeer*, but more like *tewdew, tewdew*, over and over. Last summer, when I was teaching at an island camp, I met a man who could do a perfect imitation of a killdeer. He was a fish physiologist, a quiet man who struggled to put thoughts into words. But damn, he could do birds. His meadowlark was exactly right, even with that little roll of water at the end. He could do a mouse well enough to turn a kestrel's head. I saw it happen. And his loon, my god, a loon to bring tears to your eyes, the long sad call. I watched him work with cedar waxwings in a clearing on the island, practicing the slip, the tight little whistle, until the waxwings were giving it back to him, correcting him gently, and he answered them until he had it right.

When he was a little kid, he told me, he had to get up early for his paper route. Every morning, he was alone in the dim, empty street, except for mourning doves, calling. He started to answer them, he said, and before long, the doves began to follow him, a lonely little kid weaving down the street on his bicycle, balancing a canvas bag of newspapers, leading a parade of mourning doves and cooing.

He turns away from people when he whistles, so no one ever sees him make the sound. He has to twist his face to get it right, he explained, and he's self-conscious about the shape his mouth takes on. One evening, I saw him standing alone at the end of the dock as the last of the color left the sky. I heard a loon calling, so I scanned the empty horizon with binoculars, then turned to study the broad back and averted head of a man who was crying out as if his heart would break.

At the breakfast table the next morning, I was leafing through bird books, trying to identify the owl that had called in the birches all night. "Was it like this?" the man asked, and out came a rough hooing with a gulp at the end, and that was it, exactly. "So it's the barred owl," he said, "who gives us songs in the night."

Who gives us songs in the night. I recognized the phrase from Job and thought how strange it was that a fish biologist would talk that way. Back in my office, I looked the passage up. Job 35:10. Job has been complaining to Elihu about his misfortunes, and who can blame him, having suffered, after all, the trials of Job? But Elihu says, people call for help from God, but no one praises God the Maker, who gives us songs in the night. That's the one gift Elihu wanted Job to take account of especially—the songs. "Where were you," he asked Job, "when the morning stars sang together, and all the sons of God shouted out for joy?"

Job was mortified. He had apparently not been listening, or if he heard the songs of the stars, he forgot to give thanks to God. "I lay my hand on my mouth," he wailed. "I repent in dust and ashes."

I don't claim to understand the Bible. But I wanted to know what songs meant to Job, so I carried the text down the hall to my colleague, who does. He found the passage in his copy of the Parallel Bible, a book that sets translations side by side. It turns out that *songs* is translated differently in every edition of the Bible. In the New Jerusalem Bible, it's "Who makes *glad songs* at night." And I thought, yes, this translator would understand what it's like to sleep on a beach with the frogs in full chorus and the coots hooting like drunks. The glad songs.

And in the New Revised Standard Version it's "Who gives us *strength* in the night." And I thought, this is a translator who has come onto the dunes on a night when the wind lifted sand and sent it streaming like the Milky Way over the escarpment where a mountain lion stood, watching the dunes, white under the moon.

And the Revised English Bible says, "Who gives us *protection* by night." I thought, this translator must have slept on an island in bear country, right on the soft sweet ground, right there in the salmonberries and sword ferns. He has known, from the trilling of tree frogs and the hoot of a great horned owl, that all was well. "All is well," the frog songs told him, "there are no bears afoot on this dark night."

But the New American Bible says, "Who gives us *vision* in the night." And I wondered, did this translator know the sudden seeing that has nothing to do with eyes?—that clear, sharp knowing that comes only once or twice in a lifetime, the grateful understanding, like waking up from a nightmare set on a menacingly dark street to find the sun in your eyes and nuthatches fussing in the ponderosa pines, when light has never been so luminous or colors so clear.

I wish I knew that one Hebrew word that means *vision* and *strength* and *protection* and *glad songs*. This is a word I could use. This is a word that teachers should drill in fifth-grade vocabulary lessons, the language of praise and celebration, a word I could teach to the man who sings like a cedar waxwing.

FRANK AND I STAY UP LATE IN CAMP, STIRRING A LITTLE RUM INTO our tea and talking. Bats stay out late, too, zinging past our heads. Frank's guess is the little brown myotis, but it's hard to know. They are just dark motes veering around, fluttering their wings so fast they disappear. A hunting bat sprays out a constant stream of high pulsing squeaks beyond the limit of human hearing. When a sound wave hits an insect, it bounces a signal back to the bat, who swoops in and catches the insect in a net it makes by spreading its tail between its legs. Once we were sitting beside a river in the evening. Maybe the Deschutes or the Rogue; it was a long time ago. Suddenly moths let their wings go limp and flopped spastically to the stones. Frank said a bat must be hunting; the moths felt the sound pepper their wings and took evasive action.

I ask Frank how bats keep up that constant barrage of sound. I can't imagine dropping from the ceiling of a cave, flapping my wings so fast I fly like the wind, and sending out an unending fusillade of squeaks while I'm doing it. I could fly or squeak, probably, but not both at the same time. As it turns out, he said, bats' wings work like bellows. Every time a bat forces its wings down, the muscles compress its chest and send a puff of air through the reeds in its throat. A bat works on the same principle as the organist who pumps away at the foot pedals on the organ until the organ exhales Bach.

We are surprised how long it takes before the frogs start to sing. The chorus finally starts in the marsh past our feet and spreads up the shoreline, like a wave at a basketball game, until we are awash in frog song. It's a mystery to me, how small a tree frog is and how big its song, a song so glad, so strong that a female heavy with eggs frog-strokes through threads of algae, risking large-mouth bass, risking great blue herons, unable to stay away from that song.

Our ancestors made themselves silent and small when they found themselves alone in the night. If they dreamed, they dreamed without a sound. I know I can't scream in a dream; I don't know if anyone can. Chased by dark forms darting through circles of light under streetlamps, I push air toward my lips, but it's as if my lips are sewn shut with black threads. Then, only a muffled cry pushes from the back of my throat, and dear Frank reaches a hand out of his sleeping bag and pats my head. And of course it has to be this way. If our ancestors cried out in the night, predators for half a mile around would prick up their ears and slowly turn their heads toward the source of the sound.

We sit quietly for a long time, listening to the frogs. Ground fog, which had been quietly rising from the marsh, begins to reach toward us with ragged fingers. Clouds are suddenly on the move, breaking into an armada of small ships slowly sailing past the moon. We can go into the tent or freeze, so we go to bed, leaving the tent open to the sounds of owls.

"DID YOU KNOW," FRANK ASKS, "THAT SOME OWLS HAVE UNEVEN ears?"

He has wakened me to listen to a barn owl across the lake. This happens all the time: I think I'm awake and he's asleep, but he has to wake me up to hear the songs.

If you're going to find something in the dark, he explains carefully, matched ears aren't good enough. Symmetrical ears will tell you where the prey is on a horizontal plane, if the ears are far enough apart. If a mouse is squeaking straight in front of you, the sound will be the same in both ears; but if it's off to the side, the sound will come first to the closest ear, and with a different volume and tone. Then you can swivel your head like a radar dish until you're looking straight at that invisible mouse. But how will you know if it's above or below that plane? For this, owls need asymmetrical ears. The ear hole is higher on one side than on the other. When the squeaking is equally loud in both ears, the mouse is at ear level. With its widespread, uneven ears, an owl can put a mouse right in the crosshairs of its hearing. *Binaural localization*: that's the language the scientists use, the precise words for this particular species of marvel.

Across the lake, air passes through the owl's syrinx, a bony structure at the bottom of the trachea. Specialized muscles tighten and loosen the membranes in the syrinx, tuning it like the skin of a drum. Another owl answers from the forest. Surf rumbles behind the hill of sand. A train whistle blows far away, as if someone were pressing his whole palm on the keys of an organ. Frank and I listen with our senselessly even ears. Then we stick our heads out of the tent together. Frank's ears are higher than mine, and we laugh as we both point toward the invisible owl.

Lift up your heads, and be lifted up.

SHOREBIRDS BEGIN TO CALL AT FIRST LIGHT, THEIR VOICES HIGH, piercing, and sharp. I pull warm piles of sleeping bag to my chin and lie there listening, glad to be awake again on such a morning. Spotted sandpipers hurry over broken reeds, peeping endlessly. The clouds are low on the dunes, but there is no sign of rain. A Caspian tern sails across the lily pads, screeching with a voice so sharp and tough, I feel it could scrape off the top of my head. The calls of shorebirds, which evolved at the edge of the sea, have high frequencies, audible over the low rumble of surf. In the forest, birds have low-frequency voices because the long wavelengths of the low tones are not as quickly scattered or absorbed by the tangle of leaves and moss.

Scientific research shows that birds on the floor of a jungle sing in lower voices than birds in the tops of the same trees, and the northern forests carry the basso profundo voices of the owls and the grouse. But put a bird in an open meadow or marshland where sound can carry forever in the sunlit silence: here are the voices so beautiful to the human, also a savanna creature.

AS THE SUN BRIGHTENS THE SKY BEHIND THE FOREST, I HEAR A meadowlark's slippery warble, as joyous a birdsong as there ever is across a marshy swale, and when the new sun flares under the clouds, a red-winged blackbird calls. Its song is a celebration, *okalee-ah*, and a soft *schlick, schlick*, like a wing slicing water. I love this.

I once spoke to a convention of National Park Rangers, people who love the land if anybody does—intently, pragmatically, in the rhythm of their daily lives. "I like what you say," one ranger commented afterward, "but I wonder if you couldn't say it without using the L-word." It took me a couple of beats to figure out what he was talking about, and then I didn't know how to answer. It isn't enough to point out that Hallmark has kidnapped the word *love* and beaten it senseless. My first instinct was to think that the ranger shared in the sadness of all science, its loneliness. Generations of scientists and land managers are thrilled by natural creation, but trained, like secret lovers, to a deep and steady silence about their feelings toward what they study so intently.

The physicist and essayist Chet Raymo wrote that one result of the split between spiritual and scientific views of the world is that science has lost the ability to celebrate natural creation. "In going their separate ways," Raymo wrote, "the Church and science were each impoverished, . . . and science was deprived of access to the Church's rich traditional language of praise." And more, he implies: to the extent that science has become the dominant Western worldview, we have all lost the language of praise. But I'm not sure.

I asked the park ranger, "So what word shall we use instead of love?"

He thought for a long time before he answered. "Maybe, instead, we should say 'listen to.'"

Listen to. To hear with thoughtful attention. To hold something close, to attend to it, to be astonished by it, to devote your life to its mysteries, to name it precisely, to wonder how it comes to be. To stay awake to it. To move closer to it in the wild and twittering night. To let it cover you and keep you safe. To me, listening is starting to sound a lot like love.

Of the 1,296 species of bats whose population trends have been assessed by the IUCN, a third are at risk: 24 critically endangered, 53 endangered, 104 vulnerable, and others for which full data are unavailable. In North America, white-nose fungus is a new and spreading risk, and wind turbines kill tens of thousands of bats annually.

(USGS.gov, International Union for the Conservation of Nature)

Listening for Bears

TO GET TO OUR SOUTHEAST ALASKA CABIN FROM THE MOORAGE, we walk along the edge of the cove on a trail known as Bear Alley. The trail cuts through head-high grasses and enters a damp tunnel of alders and hemlocks. This is the same path the brown bears follow to move from the abundance of sedge on the beach to the skunk cabbage in a dank lagoon. Another path follows a stream bed down steep mountain meadows to the cove. Bears have used this path for so many hundreds of years that they have worn it into a narrow, flat-bottomed channel— maybe a foot wide, a foot deep in the duff and moss, below lady ferns and Sitka spruce.

Our cabin sits in a blueberry thicket where the bear trail that descends the mountain meets the bear trail that circles the cove. There are many more bears than people on this Alaskan island, and at least a couple of bears follow these paths most days. From my bunk at the window in the cabin, I often wake up to see a bear grazing by the cove, big and calm as a cow.

Am I afraid of the bears? Yes, in fact, I am. These are Alaskan brown bears, sea-coast grizzlies grown huge and sleek on salmon. I want to live my allotted time, not find myself prematurely swatted into eternity because I got myself in the way of a bear. It doesn't matter that never, in the memory of this inlet, has anyone been so much as bruised by a bear; when I walk up the trail (usually carrying a five-gallon bucket of halibut or crabs) or down the trail (usually carrying a bucket of fish carcasses or crab shells), I am listening closely and paying careful attention.

That circle of beaten-down grass—how long ago did a bear bed down there? That pile of bear scat like ping-pong balls woven of sedge—has that been here since spring, when bears headed from their dens to the beach for a first meal? That soft puff of air—is that wind in the alders or a bear's anxious exhalation? That patch of ground pines popping out of the duff—if a bear had passed recently, wouldn't all the ground pines be uprooted and gnawed? Have hemlock needles fallen on that pile of scat? Has rain softened the edges of the paw-print in the mud? I stop to examine the bear scratching-post at the intersection of the trails. The bark on the tree is rubbed to a fine polish. I look for new patches of grizzled fur stuck in the cracks.

I scan the forest constantly, alert for a brown hump in the salmon-berries. I listen, turning my head. Disturbed bears don't roar. They might growl softly. Or clack their teeth. Most often, they make blowing noises. *Huff. Huff.* Quiet as the huffing is, as easily mistaken for water against stones, when it comes from behind a screen of alder trees, this is a sound that catches one's attention. Sometimes if I'm approaching a rise or a dense patch of berries, I'll let out a *hoo-eee*, as if I were calling hogs. But most of the time, on the trail, I sing. I sing the songs my mother taught me, whatever old campfire song comes to mind. So it's *I love to go a-wanderin' along the mountain track* that causes a bear to lift her head, look long in my direction, then with unmistakable dignity stroll off on a course diagonally away from mine, as if that was the way the bear wanted to go all along. *My knapsack on my back.*

I have decided this is not a bad way to walk in the woods—with this kind of listening, with this singing.

IF MY FATHER HAD LIVED LONG ENOUGH TO JOIN US HERE IN THE wilderness, he would be on his knees, poking a stick into a bear pile, sorting out a beetle's wing or identifying the genus and species of the undigested grass. My father was the person who taught me to pay attention. He was a naturalist for the Rocky River parks when my sisters and I were small. His job was to lead field trips on Sunday mornings along the river or into the beech-maple forest under the approach path for Cleveland Hopkins airport. Each Saturday before a field trip, he would load up the family, and off we would go.

My sisters and I were the designated Finders of Interesting Things. We lifted sheets of bark off fallen logs to find slick black salamanders and turned stones along the river, looking for dragonfly larvae. I remember finding a hummingbird's nest made of lichen and spider webs, tiny as an eggcup. If we found a plant we couldn't identify, we called my father over and he helped us key it out, step by step. *Jewelweed*, he would say. *Squeeze the seedpod just this way to see it pop.* We sang along with birdcalls. *Old Sam Peabody Peabody Peabody*: the white-throated sparrow. We poked sticks down crayfish mud-towers and tried to find fossils in the shale. In cups folded from mayapple leaves, we carried water striders into a puddle to watch them fight.

By the next day, my father had become a magician. With his pet raccoon on his shoulder, he led a string of people who had skipped church for a "nature walk" in the very place we had explored the day before. Look here, he would say, you can often find salamanders under the bark of rotting logs. And there it would be, skinny, shiny, glorious. Can you find a bird's nest in this fir, he asked, and someone did. Not just any bird's nest; a hummingbird nest made of lichen and spider webs. He knew every bird call, knew the name of every plant, in Latin and in English. He knew why only female mosquitoes bite. He knew why stinging nettles sting. It didn't matter that airliners regularly roared over the tops of the trees or that teenagers washed their Thunderbirds in the river upstream. For the people on the field trip, the morning was a great satisfaction. They found what they were looking for, which, as I think about it now, must have been an intimacy with the everyday marvelous, the miracle you can cup in your hand.

If my mother were here walking the bear trails, she would be organizing us to sing rounds. You start, you're second, you come in third, she would say. If we are going to sing to bears, we will give them a rousing chorus, not insult them with a single line of song. It is harmony, after all, that we seek. Harmony, that will save us. *All things shall perish from under the sky. Music alone shall live, never to die.* If my mother were here, we would march across the mudflat, keeping one eye on the bear on the far side of the cove, making harmony the way a round makes harmony, weaving the song like sedge is woven in bear droppings in the spring, pausing to hold an especially beautiful chord as a gift to the bears. My mother taught us dozens of rounds, and so we learned to listen, to be attentive to the music of others, to tune ourselves to their chords, to pace ourselves to their rhythms. We learned that weaving songs is one of the most beautiful things we can do, and that we can't do it by ourselves.

My mother was a Girl Scout when she was young, just come from England, and that's where she learned the campfire songs that yearn for wild places. *It's the far northland that's a-calling me away / as I take me with my packsack to the north.* That was before she went off during World War II to help rehabilitate soldiers. She was the nurse who led the singing therapies. I can imagine all the grievously wounded soldiers, pale on white sheets, singing *I love to go a-wandering, along the mountain track.* And maybe that was healing. Maybe that was exactly what they needed, to sing in harmony from all the beds in the ward, to sing in harmony, from the place of their lonely grief, the song of the "Happy Wanderer."

CAN IT SAVE OUR LIVES, THIS ATTENTION?

In a literal sense, it might someday. Listen closely, and you might hear a bear before it hears you. Then you can stop, group up with your friends, and find another way to go, even if it means crashing through the devil's club that stings the backs of your hands. Or sing, and the bear will hear you coming even if the creek is in her ears. Then, maybe she will remember she has an appointment in another place.

So attentive listening might save someone in that small way. But there's much more to it than that. Thomas Berry wrote:

> We are most ourselves when we are most intimate with the rivers and the mountains and woodlands, with the sun and the moon and the stars in the heavens; when we are most intimate with the air we breathe, the Earth that supports us, . . . with the meadows in bloom. . . . However we think of eternity, it can only be an aspect of the present.

Each of us is so much more than we think we are—this body, these sorrows and hopes. We are air exhaled by hemlocks, we are water plowed by whales, we are energy ejected from stars, we are children of deep time. Our ears tremble with wind through treetops. Our eyes flash with sunlight through rain. How can we be fully alive, if we don't pause to notice, and to celebrate, all the dimensions of our being, its length and its depth and its movement through time?

Here on the island, our family takes great trouble to extend the *length* of our lives. The first thing visitors notice is the bear spray in the outhouse. The second is the shotgun on the shelf above the cabin door. What if we put as much energy into extending the *depth* of our intimate connections to each moment we live? I think this is what listening for bears can do. That sort of attention alerts us to so much more than bears.

Musician Pauline Oliveros recommends, "Take a walk at night. Walk so silently that the bottoms of your feet become ears." In bear country, this is terrible advice. Bears walk at night. They walk so silently that the bottoms of their paws become ears. But yes, I get it. The profound listening she describes is a way to hear, and so to gather meaning that deepens an intimate connection to the Earth. Listen to a thrush and hear the eons in the evolution of a birdsong. Ring one stone against another and attend to the layers of time embedded in the boulder on the beach. Splinter a spruce stump with a splitting maul and hear the successive years fall away. Hear the surf in the stones and wonder at how stones and wind-whipped willows came to sing in unison. Hear their music and wonder why it rings such a deep chord in you. In entering into Earth's enduring song, we come as close to eternity as we will ever be.

Although *Ursus horribilus*, the brown or grizzly bear, once roamed the North American west, the 300,000 remaining bears are found primarily in the mountains of Canada, Alaska, and, to a sharply diminishing extent, the lower forty-eight states. The International Union for the Conservation of Nature lists the brown bear as a species of least concern, although grizzly bears in the Yellowstone area are listed as threatened.

(International Union for the Conservation of Nature, Wikipedia, Sierra)

2.

WEEP

LISTEN: THE ALBATROSS IS WHISPERING TO HER EGG ON THE nest. Her long beak, tracked with white trails of the salty tears she has shed, reaches down to touch the shell that holds her child. Small sounds, *whistle-tick, mumble*—surely the blind, curled chick hears his mother. Does she pray? a woman asks. *May seas be kind, may anchovies be plentiful, may winds be steady under your wings.* Does she promise? asks another. *I will nudge you into the sweet feathers of my chest and sing to you, a whistling song. Until you fly from me forever, I will keep you safe and defend you steadfastly.* A third woman asks, Does she, in her sorrowful wisdom, warn the chick? *Soon, you will spread your wings and soar into a world gone suddenly cruel, the winds ferocious and cross-wise, the sea currents confused, the fishermen desperate, the sun hot on the seas. The ocean will offer objects that will not nourish you—cigarette lighter, toothbrush, bottle cap, button, bread bag, baited hook, doll's severed arm—but sardines and bloodied herring, torn by feeding whales, will be few. You will be hungry.* Does she say that? Does she silently weep?

The Tadpole Motet

11:30 A.M. AS I STUDY THE PATTERN OF RAIN ON THE POND, THE surface breaks open and a hooded merganser pops out. As soon as my binoculars bring her into focus, she shrugs—*hup!*—and dives. I presume she is winging around down there, slurping up tadpoles. It is early April, and the pond is thick with them, black dots with frantically wriggling tails.

Until the squall moved in, I had watched violet-green swallows swing over the water, snatching mayflies or midges or mosquitos, I don't know exactly which. They all look like glitter, ascending. I can understand how a swallow could catch a crane fly as it slouches along with dangling legs. But the timing required to snatch a single gnat from a cloud of gnats? That's beyond imagining.

And here's an even greater mystery of timing: How does the swallow, fattening up over salt marshes in Mexico, decide that she must leave exactly that day if she's going to arrive at the pond in Oregon just as the insects are emerging? And how does she pull off that miracle of foresight, to lay her eggs so they will hatch just when the caddis flies are crawling up cattails into the light? To a swallow, timing is everything.

Scientists are quite sure that the swallow's migratory restlessness is triggered by the length of the day, and that the insect emergence is triggered by the temperature. But scientists have little to say about the nature of the conductor whose waving hands mark time in this planetary motet, cuing light, then warmth, now the migrating ducks, now the greening weeds, the budding trees, the returning rains, the tadpoles and the gnats, the squalls blowing in from the sea. Of course, there are always stutters and starvation. But generally, the timing has been close enough, and the animals evolved in the dependable rhythms of this world, living and dying to the drumbeat and the hum.

The Earth still slowly spins; the cycling days remain steadfast. But weather now comes and goes with wild and violent variations, and we all struggle to make sense of it. It used to be that no one blamed the weather, because we knew it had no will, no viciousness or injustice. But the atmosphere that generates the weather has lost its innocence. The hand that now conducts the swirling chorus of air and water is the human hand of commerce, flinging greenhouse gases into the air, altering the currents of wind and water, disrupting the rhythms in the ancient symphonies of small lives.

Last year, the swallows got the timing wrong. They came back to Oregon before winter was finished, and there were no insects in the wind. I have seen a starved swallow, its wing frozen to the sand. I have seen the frozen eye of a swallow. It's white. You can't see into it.

This year, I found a dead bat in the corner of my cabin, where a slat had fallen loose. He was a tiny thing. His black wings were extended, his face locked in a bat's toothy grin. I was sorry to see it dead; I worried that the solar gain through my window had tricked the bat into waking early, into thinking it was spring when it was still winter, and it starved. To a bat, too, timing is everything. Now midges spin in small cyclones over the pond and mayflies rise from bent reeds. But in the air rising from the pond tonight, there will be no little bat with moonlight in its wings.

1:30 P.M. I'M STARTLED TO SEE A TURKEY FLUSH OUT OF THE grass, flapping madly. He has no sooner scrabbled onto a pine branch than a coyote trots onto the dike. The merganser runs across the water and arrows off to the northeast. I rush onto the porch with binoculars. None of this fuss alters the coyote's pace: one step, a pause to look behind, another step. Wind riffles the water and lifts the coyote's tawny fur.

The only denizen of the pond who is sitting still in all this ruckus is Frank. In a folding chair under a black umbrella, he leans toward the water, wearing headphones, holding a recorder in one hand and a bagel in the other. He wants to record a red-legged frog, a rare frog that burbles underwater. But it may be too late in the season or too early in the day. The only frogs singing now are Pacific tree frogs, the tiny green ones, but their hearts aren't in it. Most of them have already mated and laid their eggs.

The warm rains of the waning winter drew the tree frogs into the pond, and then the yelling was so deafening that the neighbors complained about the noise. That's when we brought our grandsons out to my writer's shack in the country, to probe the darkness with flashlights; to fill their boots with pond water; in every pondside hole, to find tiny gold eyes staring at them in reciprocal astonishment. To fall asleep to the frogs' lullaby, but in fact not falling asleep until a blanket of cold quieted the frogs. The boys were out the door first thing the next morning, their yellow raincoats bright against gray wind-wrinkles on the pond. "I found a newt," one shouted, and two damp heads bent low to the water. "Nonni, come see," and of course I did.

It's been much cooler since then, which has my husband worried. After the first warm rains and that orgy of singing, the weather has turned cold. Nasty cold. Now in April, it is long past time for the rainbow showers of spring to arrive, but still we have the dead-earnest rain of winter. Incessant rain loosened the roots of the big Douglas-fir across the pond, and in one last gust, it toppled, all one hundred years of it, all one hundred fifty feet and an osprey nest. The cold is going to slow the frogs down, and who knows how quickly summer will come.

Our pond is a seasonal pond, which means that every summer it dries to a shallow scoop of hardened mud and a lattice of dried pondweed. This means that the frog eggs are in a race against time from the moment they are laid. As the pond begins to sink into its muddy bottom, the tadpoles grow outsized hind legs and little *T. rex*–like forearms. Then they resorb their tails and, if they get the timing right, crawl out of the pond just before it vanishes. Larval newts—no bigger than crickets, with lacy gills and legs thin as threads—hang near the bottom, seemingly bored. But they too are in a desperate race to resorb their gills and grow big enough to crawl out of the pond before it dries. They too must become creatures of the damp land before the dry summer season comes, plodding through the last of winter's rainy days into the forest and the river.

Every year, the question is: Will the pond creatures mature fast enough to escape the pond before summer's drought? This depends on the weather, and it cuts both ways. The warmer the weather, the faster the eggs develop into tadpoles and the faster the tadpoles develop into frogs. But it's also true that the warmer the weather, the faster the pond shrinks. The froglets crowd into smaller and smaller spaces, wriggling now with their backs out of the water. They might not know that the pond is shrinking, but they can sense they're getting crowded. Perception of density triggers even faster, now frantic, rates of growth—if there is enough to eat. Frank says that the tadpole of the spadefoot toad grows a tooth when the pond gets too crowded. Then he transmogrifies into a cannibal, stabbing that awful tooth into his sisters.

3:05 P.M. THE SQUALL HAS BLOWN OVER AND THE SWALLOWS ARE back, chasing flecks of sunlight. Summer came early last year, the sun hot for weeks on end. As the water level sank, the pond became a minestrone of little wriggling things, not yet terrestrial creatures, not quite yet. We checked every day, despairing.

Too soon, the pond was reduced to a skim of algae-choked water shimmering with lives. The sun rose hard and white every morning, sank hard and red every evening. The pond shone like gold plate, wrinkled by the struggling animals. The water warmed to the temperature of urine. The pond continued to shrink.

We gathered nets and buckets. With the grandchildren, we shuffled through the hot water, scooping up small lives, lowering them into buckets and carrying them to the river. And then again. So many hopeless tadpoles, so many larval newts, impossibly delicate. They were so concentrated in the shrinking pond that none of us could walk without crushing soft bodies into the mud. The little boys were horrified, paralyzed in place, waving their nets and calling for help so they could walk without killing. But we couldn't help them without ourselves crushing the tadpoles. We did our best. We poured bucket after bucket of quivering black things into the river, and how did they survive that plunge? The boys stood ankle-deep in warm water, socks sagging around their ankles, and stared into the buckets at the tadpoles.

"Hurry," they called out. "They're dying." But when the number of crushed animals was greater than the number of animals that remained, we gave up. There was no saving them.

We found a place to sit in the shade and rinsed the mud and corpses off our sneakers. The boys lay on their backs with their arms over their eyes. "I think we saved enough just in time, do you think we saved enough, yes?" the littlest one said. "Maybe or maybe not," answered his brother.

Two days after our rescue attempt, the water was gone from the pond, and the mud was painted with a black layer, like melted tar. That tarry layer was the bodies of tadpoles and larval newts who lost the race against the uncaring heat. By the next week, the tar had cracked and curled at the edges, and who could imagine that such a dark thing could ever have grown to fill the evening air with song?

Under the Endangered Species Act, the California red-legged frog is listed as "threatened." The primary causes of decline are the fragmentation and destruction of its marshland habitats by mining, cattle grazing, urban and suburban development, and water diversions and impoundments. Other causes include spreading fungal diseases and introduced predators, primarily bullfrogs.

(U.S. Fish and Wildlife Service)

The Silence of the Humpback Whale

I REMEMBER *SONGS OF THE HUMPBACK WHALE*. THIS WAS THE SEN-sational 1970 recording of humpback whale song that brought whales into the hearts of people around the world. As the whales courted in Hawaiian bays, their plaints were almost operatic in their drama, their love lust, the lyricism of their songs. Friends gave us the LP when our daughter was born, so we could rock her to sleep to the sweet whispers and whoops of the whales. As graduate students just moved to town with a new baby, we had nothing in the house but a mattress on the floor, a record player, and a load of firewood for the woodstove. The forest smell of the damp oak, the music of the whales, the warm, gently breathing weight of a new baby on my chest—this was what the world was created to be, I believed, nothing less or more. The baby slept soundly, dreaming maybe of rising and falling on a gentle swell, lulled by the love songs of the great whales.

Forty years later, we bought a cabin on the edge of a cove in Southeast Alaska. After a long dusk that first day, the sun finally dropped below the mountains, leaving a pink glaze on the water. We slept to the wash of waves in the rock wrack. But not for long. A sudden call jolted us awake, a long, drawn-out squeal. *Did you hear that? What in god's name?* A wolf howling? It might have been, but there were no wolves on the island and the sound was chestier than wolves. An elephant trumpeting? That's what it sounded like, but no mastodons had stomped these beaches for ten thousand years. Nothing we had ever heard matched the magnitude of that bleating. A ruckus of thunks and splashings sounded from the inlet, and then the night returned to its gentle swash. In the morning, we saw a distant pod of humpbacks, spouting clouds of sunlight.

Those, we learned, were the calls of the humpback whale.

Although they are probably the very same whales that sing in Hawaii, the humpbacks of Southeast Alaska add a different call to their repertoire when they migrate back to northern feeding grounds. They are all violin music in the Hawaiian bays, but on the feeding grounds in Alaska, whales trumpet. The cacophony is part of their raucous feeding ritual, unique to Southeast Alaska. An assigned member of a pod circles deep, blowing bubbles the size of beachballs. The bubbles form a sort of cylinder, encircling a school of herring. Other whales swim below, herding the herring into a tight ball. A whale sounds the signal, that magnificent screech, and all the whales drive powerfully upward through the panicked fish, jaws agape. They go so fast, they breach the surface, sailing half a body's length into the sky. Water streams from the baleen curtains that hold the herring in their maws. Gulls scream as whales fall back onto the water with all the weight and grace of a school bus falling off a cliff.

When we are out fishing, we usually hear exhaling whales before we see the cloud of breath. One returning whale in our inlet rasped heavily every time he inhaled or exhaled. People could identify him from miles across the water. "Growler," they called him. Most of the whales exhaled in long breaths that sounded exactly like someone was dragging an ice chest across the deck of a boat. But the most beautifully breathing whales were the silent ones in fog on the far side of the inlet. When they exhaled, a cloud of silver glitter formed over their curled backs and silently disappeared. One morning in Freshwater Bay, we glimpsed a whale that was sleeping, a big lump floating so close to the surface that we were glad not to have hit it. The whale's great bulk rose with the inhalation, sank on the exhale, quietly, slowly snoring on the swell.

This is the music of the humpback whales in Alaska.

But here is what I need to tell you. The humpback whale population in Southeast Alaska had been abundant and growing at about 5 percent annually. Now, suddenly, the numbers are down nearly 60 percent over a five-year period. Lots of things happened in that time. A perfect storm of ocean events shifted prey availability and quality: powerful El Niño conditions, an unprecedented "blob" of warm water in the Gulf of Alaska, global warming, harmful algal blooms. A concurrent mass die-off of seabirds signaled widespread prey shortages. Whales in Southeast Alaska were visibly thin, and even the zooplankton were skinny, measuring lower levels of lipids. Glacier Bay and adjacent waters in Icy Strait usually nurture about ten new humpback calves every summer. Last year, there was one baby, and it disappeared. Most likely it died and, too thin to float, sank to the bottom of the bay. Imagine the music of that dead baby whale, the scurrying crabs and clicking shrimp, the swish of hagfish, the rasp of shark skin against the small flayed body.

No one knows if the whales have shifted feeding grounds to follow dwindling bait fish, or if they have died. If that many humpbacks have died, one would expect a plague of dead whales washing up on beaches, but there have been none. That might make sense; an emaciated whale may sink quickly in cold water, and then the pressures of the deep sea may hold the carcass on the sea bottom, a banquet for the hungry ocean. Another plausible scenario is that whales, unable to store enough fat on the feeding grounds, set off for Hawaii nonetheless, and don't make it. Whale numbers are down in Hawaii as well as up north. No one knows what will happen to the humpback populations next, but the trends—the rising temperatures of the water, the falling populations of feed-fish and zooplankton—draw a jagged falling line on graphs.

WHAT EXACTLY WOULD BE THE NATURE OF THE WRONG, IF WE were to let whale-song slip away or, worse, propel it into oblivion? There are a number of words to use, human beings being prodigious inventors of varieties of wrongdoing. If Inuit people have forty words for *snow*, as I am told, how many words does the western world have for *wrong*? I can think of five big ones. Tragedy. Injustice. Profanity. Cruelty. Disrespect.

1. Tragedy. Let's start with this, although we won't linger. When I look out my window now, the inlet is flat as a silver plate, dinged here and there by a merganser or loon. I watch for whale spouts. Although I can see five miles across the inlet and even farther in both directions, I do not find them. That is a true loss. Seeing whales makes me glad. So large (a floating school bus), so mysterious in their underwater travels (the great migrations), so ponderously clever in their lifeways (the underwater nursing calves), so beautiful in their shining dives (the waterfalls from lifting flukes), so oddly wonderful (the stalked eyeballs that allow them to see into their own mouths), so full of life (the triumphant roar). They lift me out of myself and invite me into something far greater than my paltry concerns, into the infinitude of evolution and the great mysteries of beautiful life. Just to be in view of that is a joy, and when once I had the chance to move in close to a whale and breathe in the whale's exhalation, I was beyond happy (even after I learned about the bacteria in the exhaled breath).

I'm not alone. In our inlet, tourists on the *Island Song* line the rails in bright raincoats, holding long-lensed cameras. They cheer when a whale spouts, an excitement we can hear a mile across the water. The scene makes me think of photos of sailors returning to port after the war—that eager, that glad, that crowded at the ship's rail. This is a mystery in itself, why humans are drawn so strongly to the great mammals, as strongly as they might be drawn to their home after a war. But it seems to be so.

This joy is part of the instrumental value of whales, their worth as a means to human ends. It is a value—but utterly egocentric and insulting, when you think about it, to imagine that the value of the whales is primarily their value to us. Imagine the long evolutionary journey of whales, dragging themselves onto the muddy shore, stalking the swamps on doglike legs, swinging elongated heads, and then finally splashing back into salty water, their feet sucking mud, their mouthparts maybe mumbling like crabs, the air electric with thunderstorms and erupting volcanoes. Imagine the slow movement of their nostrils to the top of their backs and the transformation of a tail into those splendid flukes, black tulips of the sea. Imagine the evolution of that hulking grace. And where did the baleen come from, and over how many million years, the feathery filters stuck with windrows of krill? And the songs: How many generations taught how many generations to sing songs so compelling that they outsold the Monkees? To what end? That I would smile at night to hear them howl? That's all?

Let us grant the terrible sadness we would feel if the whales disappeared. Let us grant the tragic unfolding of human folly. But let's reason past our own selfish interests. Apart from these, what exactly is wrong with letting whales slip into oblivion?

2. Injustice. With the whales and all of Earth's beings, we share the kinship of common substance, the kinship of common origins, the kinship of interdependence, and—perhaps disastrously—the kinship of a common fate. There are no natural hierarchies of deserving in this planetwide family. If we and whales have evolved as interdependent and equally remarkable parts of a morally worthy whole, then we acknowledge also the *moral* unity of all life. So, a planetary argument by analogy unfolds: Just as humans ascribe intrinsic value to themselves, value beyond their usefulness to others, so the rest of creation too has intrinsic value. Just as humans grant legal and

moral consideration to their own interests, so the interests of all others are worthy of consideration. And just as humans grant themselves rights that protect their most necessary interests, so the rest of creation too has the right to protection of their essential interests.

Industrial-age humans have been slow to realize that all members of the Earth community have rights. Steeped in self-glorifying narratives of human superiority over the rest of the natural world, intoxicated by seemingly limitless power to turn nature to human uses, blinkered by short-term self-interest, humans have chosen to reserve rights for themselves. However, the narrative of human exceptionalism is increasingly challenged by a notable convergence of religious, Indigenous, ecological, and evolutionary insights. We understand now that not only human beings but other living beings, species, ecological communities, landscape formations, and waters have interconnected *interests*. Humans are morally obliged to recognize and to weigh these interests in decisions that impact nature. That is to say, other-than-human members of the Earth community also have *rights*, and those rights *count*.

Accordingly, the Universal Declaration of the Rights of Mother Earth and other legal and moral documents around the world encode nature's rights as a "common standard of achievement." The rights include, among many others, the right to life and to exist; the right to regenerate its biocapacity and to continue its vital cycles and processes free from human disruptions; the right to be free from contamination, pollution, and toxic or radioactive waste; and the right to maintain its identity and integrity as a distinct, self-regulating, and interrelated being. That means whales.

Damage to the whales—whether by overfishing their food species, acidifying the very water they swim in, degrading the zooplankton they feed on, warming the water (the list is long)—violates their rights. And it is a particularly pernicious violation, because the whales are the very definition of innocent, having done nothing to deserve this cruelty.

3. Profanity. Let me tell you about one day a dozen years ago, a special day but not a unique day. The whales had been feeding in the inlet, but they were resting now on the glaze of the sea, and our boat rested some distance away. There were many whales. They all sucked bright day into their lungs, blew it out with the sound of a rockslide. Then there was silence except for the whispers of murrelets and the flicking fins of wounded fish. Already, the sea had melted the rough water, skinning it with silver. Gulls swayed on the swell, and even the sacrilegious gulls were silent.

A whale folded its back, slowly unfolded, and levered its flukes into the air. The tail stood like a black jib, streaming water, then sank as the whale dove to a seam below the reach of the sun. Water slipped into the space the whale had pressed on the sea. One by one, other whales raised their flukes and dove. The gulls, still silent, waited. They knew that in their own time, the whales would begin the hunt again. The water rose and fell in meditative breath.

I don't want to say that the moment felt like church, because I don't want to default to human comparisons, but it felt somehow sanctified. That moment, and those whales, were beautiful and fearsome, beyond human understanding. If this is the language of the sacred, then let us use those words. This is the sanctity that we must protect, the endlessly creative world that we must save, the lyric voices that we must hear, the wonder that we must preserve.

Every extinction, every suffering, every destruction is a profanity, a failure of reverence. It is a violence we cannot even begin to measure because we have only the sorriest understanding of the world's multitude of lives. The world is a mystery of infinite and intrinsic value.

4. Cruelty. None of us can directly experience the pain or sorrow felt by another creature. We infer it from cries and pleas, and from analogy to what we ourselves would feel. The sorrow of a mother whale, faith-

fully nursing her calf through the watery nights, but too starved herself to provide the nutrients to keep the little one alive: What agony is this? It might be less than you or I would feel, but it might as likely be more, the breaking of a great whale's heart. One might argue that a whale doesn't have the mental capacity, the consciousness or self-awareness, to grieve. One might argue that she doesn't remember pain, a merciful amnesia. But these would be arguments from ignorance. We just don't know. But we can imagine.

If there are any limits to permissible human behavior, then surely cruelty to innocent creatures is beyond the pale. Pain inflicted as an unseen and unintended consequence of activities aimed at other, maybe admirable, goals; pain inflicted as a foreseen but discounted consequence of other activities; pain inflicted knowingly and intentionally as part of a business plan to drive up corporate profits—here is an escalating scale of shameful behavior. As we think about the extinction crisis, as we count down the numbers, as we calculate the rate of ecosystem collapse, it is essential to remember that the crisis shimmers with suffering. That makes it not just an environmental crisis but a moral catastrophe.

5. Disrespect. A great whale is a wondrous creature, so tuned to the flashing fish and the dark sea. It is beautiful, the glistening blue-black back decorated with barnacles, studded with scars from cookie-cutter sharks, a mammoth animal as graceful as flowing water. A great whale is knowing, as elders are knowing, having seen the world's cruelty and promise. It is magnificent beyond human measure, slowly folding and unfolding through time. It is roaring grand. It is eager for ongoing life. It is a trembling consciousness, a manifestation of the mind of the universe.

It is worthy. That's the word. It is excellent. And so it must continue. And the thought that we humans might trade the humpback whale, for what? The profligate burning of oil and gas? Profits from a reckless herring fishery? A failure to imagine a sustainable way to live on Earth? Greed, pure and simple? That is moral monstrosity on a cosmic scale. It's time for a new global conversation about the true worth of the world's great diversity of lives, not in the pinched terms of human financial or emotional neediness, but in terms of the "great journey of the universe" toward an abundance of ongoing life.

LAST YEAR, UNDER GATHERING CLOUDS, I KNELT BESIDE A TIDE-pool. Maybe you have done the same. Blue mussels paved the rocks, cutting my hand when I turned a stone. The bottom of the stone was slathered with life—tiny starfish, algae like orange paint, crust-of-bread sponges, porcelain crabs disguised as pebbles, decorator crabs disguised as seaweed, fish disguised as rays of light. The moving tide was noisy, the harsh inhale and groan. Scritching claws and bubbling jaws, a constant *plop plop* as seawater dripped off globules and tentacles and who knows what. Behind me, I could hear my grandsons calling to each other, "Guys! Come. Look and see." And then, out in the inlet, a humpback whale began to roar.

Never have I heard as complete a repudiation of the idea that human beings are the only wondrous beings, that we are in charge, that we are the point of the whole thing. Each being is worthy. Each fractal layer is necessary, all the lives the theme, all the lives the variations. The planet is still crammed with lives of urgent striving, crawling over each other, burrowing into every crack, floating on the seas. The fate of these lives is not a matter of indifference or of economic expediency. These lives are the irreplaceable consequence of planetary creativity over 4 billion years. As consequences of the same creativity, we human beings have obligations to honor the Earth's beings and the processes that created them, to celebrate and protect them until the end of time.

In the summer habitats of humpback whales in Southeast Alaska, researchers are finding increasing signs of stress. The numbers of whales returning from breeding grounds in Hawaii have sharply declined, and those returning are visibly thinner than they usually are at the end of their feeding season. The number of calves has declined, as their mortality rates have increased.

(National Oceanic and Atmospheric Administration)

The Meadowlark's
Broken Song

THE SOUTHERN TIER OF SASKATCHEWAN IS FARMLAND AND prairie—dry, wind-raked grassland with a few fence posts and barbed-wire strands to mark boundaries. I don't want to say the land is pristine. Only 1 percent of the tall-grass prairie remains, 18 percent of short-grass prairie. Much of the rest has been converted to the production of grains. But the fracking pads, their roaring flares and pounding pumpjacks, are generally to the north. So the Canadian meadowlarks have basically what they need, which is open space, seeds and grains, bugs, beetles, and now and then a dip in the ground, a cow's footprint maybe, where the meadowlark can build her nest of stalks and grass. Most importantly, they have quiet.

Quiet is essential, because nestling meadowlarks learn their flute-like songs from listening closely to their parents. In the quiet prairies of Saskatchewan, the song of the meadowlark is complex and nuanced. "Elaborate, bubbly, beautiful" is how Gordon Hempton, an acoustic ecologist who specializes in recording natural sounds, described it to me. The song was so elegantly baroque that even he didn't recognize the species at first.

South of the Canadian borderland prairies, however, in northeast Montana and in North Dakota, fracking drill rigs are so thick that, standing at one, you can always see another in the distance. You can always hear their din. With noise that can be heard for ten miles in all directions, the racket overlaps and multiplies, burying the prairie in noise. Drills screech metal on metal, pumpjacks scrape and pound, compressors roar like jet planes, huge trucks roar by, and the rock howls nonstop, twenty-four hours a day. While the sound of a typical neighborhood is 35 decibels, the sound of a fracking field is 83. Decibels are measured on a logarithmic scale, so an increase of ten means the sound is ten times louder. The sound of a fracking field is roughly comparable to a garbage disposal in the kitchen or a diesel train going 45 mph a hundred feet away.

Because there is no quiet, young meadowlarks cannot hear their parents sing. As a result, when they become adults, their songs are broken—attenuated and simplified. Gordon compares the stunted language of the meadowlarks to the language of the battlefield, where commands have evolved to be short, simple, loud, and entirely without nuance, to be heard over machine-gun fire and exploding artillery shells. For centuries, across the sea of bison-studded prairie, meadowlarks have used their songs to establish territories, seduce mates, and—who knows—maybe to shout for joy. The loss of a beautiful song is the first installment on the terrible price they pay for oil.

TWELVE YEARS AGO, I WROTE AN ESSAY THAT PEERED AHEAD IN time to envision a world that had lost many of its songs—the frogs, the bats, splashing salmon, meadowlarks. It broke my heart to write it, imagining what those losses would mean to the children. In that essay, I held my newborn granddaughter asleep in my arms as I looked ahead, sad and scared.

My granddaughter is twelve years old now. She lives in Canada. Both of her parents are biologists. She knows where to find the nest of a barred owl. She knows how to hold a frog without hurting it. She plays keyboard in a rock band and alto clarinet in the concert band. She has a beautiful singing voice and will sing you a song about fireflies. But she has never seen a firefly. And she has never heard a meadowlark sing.

So here is that essay again, for Zoey. I have restored some of the details that, in a small act of mercy, I censored from the first version. It's called "The Angelus," which is the call to forgiveness at the end of the day.

ALL THOSE YEARS, THE SWAINSON'S THRUSHES WERE THE FIRST to call in the mornings. Their songs spiraled like mist from the swale to the pink sky. That's when I would take a cup of tea and walk into the meadow. Swallows sat on the highest perches, whispering as they waited for light to stream onto the pond. Then they sailed through the midges, scattering motes of wing-light. Chipping sparrows buzzed like sewing machines as soon as the sun lit the Douglas-firs. If I kissed the knuckle of my thumb, they came closer and trilled again.

For years there were flocks of goldfinches. After my husband and I dug out the bull-thistles and teasels on the far side of the pond, the goldfinches perched in the willows. When they landed there, dew shook from the branches into the pond, throwing light into new leaves where chickadees chirped. The garbage truck backed down the lane, beeping its backup call, making the frogs sing, even in the day.

Oh, there was music in the mornings, all those years. In the overture to the day, each bird added its call until the morning was an ecstasy of music that faded only when the diesel pumps kicked on to pull water from the stream to the neighbor's Bing cherry trees.

Evenings were glorious too. Just as the sun set, little brown bats began to fly. If a bat swooped close, I heard its tiny sonar chirps, just at the highest reach of my hearing. Each downward flitter of its wings squeezed its lungs and pumped out another chirp, the way a pump-organ exhales Bach. Frogs sang and sang, but not like bats or birds. Like violins, violin strings just touched by the bow, the bow touching and withdrawing. They sang all evening, thousands of violins, and into the night. They sang while crows flew into the oaks and settled their wings, while garter snakes, their stomachs extended with frogs, crawled finally under the fallen bark of the oaks and stretched their lengths against cold ground.

I don't know how many frogs there were in the pond then. Thousands. Tens of thousands. Clumps of eggs like eyeballs in aspic. Neighborhood children poked them with sticks to watch their jelly shake. When the eggs hatched, there were tadpoles. I have seen the shallow edge of the pond black with wiggling tadpoles. There were that many, each with a song growing inside it and tiny black legs poking out behind. Just at dusk, a hooded merganser would sweep over the water, or a pair of geese, silencing the frogs. Then it was the violins again, and geese muttering.

In the years when the frog choruses began to fade, scientists said it was a fungus, a virus, or maybe bullfrogs eating the tadpoles. No one knew what to do about the fungus, but people tried to stop the bullfrogs. Standing on the dike, my neighbor shot frogs with a pellet gun, embedding silver BBs in their heads, a dozen holes, until she said, "How many holes can I make in a frog's face before it dies? Give me something more powerful." So she took a shotgun and filled the bullfrogs with buckshot until, legs snapped, faces caved in, they slowly sank away. Ravens belled from the top of the oak.

When the bats stopped coming, they said that was a fungus too. When the goldfinches came in pairs, not flocks, we told each other the flocks must be feeding in a neighbor's field. No one could guess where the thrushes had gone.

Two springs later, there were drifts of tiny white skins scattered in the shallows like dust-rags in the dusk. I scooped one up with a stick. It was a frog skin, a perfect empty sack, white, intact, but with no frog inside. It has been cleaned, I supposed, by snails or winter. There were dozens of them, empty frogs scattered on the muddy bottom of the pond. They were as empty as the perfect emptiness of a bell, the perfectly shaped absence ringing the angelus, the evening song, the call for forgiveness at the end of the day.

As it happened, that was the spring when our granddaughter was born. I brought her to the pond so she could feel the comfort I had known there for so many years. Killdeer waddled in the mud by the shore, but even then, not so many as before. By then, the pond had sunk into its warm, weedy places, leaving an expanse of cracked earth. Ahead of the coming heat, butterflies fed in the mud between the cracks, unrolling their tongues to touch salty soil.

I held my granddaughter in my arms and sang to her then, an old lullaby that made her soften like wax in a flame, molding her little body to my bones. *Hush a bye, don't you cry. Go to sleep you little baby. Birds and the butterflies, fly through the land.* I held her close, weighing the chances of the birds and the butterflies. She fell asleep in my arms, unafraid.

I will tell you, I was so afraid.

Poets warned us, writing of "the heartbreaking beauty that will remain when there is no heart to break for it." But what if it is worse than that? What if it's the heartbroken children who remain in a world without beauty? How will they find solace in a world without wild music? How will they thrive without green hills edged with oaks? How will they forgive us for letting frog song slip away? When my granddaughter looks back at me, I will be on my knees, begging her to say I did all I could.

I didn't do all I could have done.

It isn't enough to love a child and wish her well. It isn't enough to open my heart to a bird-graced morning. Can I claim to love a morning, if I don't protect what creates its beauty? Can I claim to love a child, if I don't use all the power of my beating heart to preserve a world that nourishes children's joy? Loving is not a kind of *la-de-da*. Loving is a sacred trust. To love is to affirm the absolute worth of what you love and to pledge your life to its thriving, to protect it fiercely and faithfully, for all time.

My husband and I were there when the last salmon died in the stream. When we came upon her in the creek, her flank was torn and moldy. She had already poured the rich, red life from her muscles into her hopeless eggs. She floated downstream with the current, twitching when I pushed her with a stick to turn her upstream again. Sometimes her jaws gaped, still trying to move water over her gills. Sometimes she tried to swim. But she bumped against rocks, spilling eggs onto the stones. Without reason, she pushed her head into the air and gasped. We waded beside her until she died. When she was dead, she floated with her tail just above the surface, washing downstream until she lodged on a gravel bar. The music she made was the riffle of rib bones raking water, then no sound at all as her body settled to the bottom of the pool.

I buried my face in my hands, even as I stood in the water with the current shining against my shins. Oh, we had known the music of salmon moving upstream. When the streams were full of salmon, crows called again and again, and seagulls coughed on the gravel bars. Orioles sang, their heads thrown back with singing. Eagles clattered. Wading upstream, we walked through waves of carrion flies, which lifted off the carcasses to swarm in our faces, buzzing like electrical current. Water lifted and splashed, swept by strong gray tails, and pebbles rolled downstream. It was a crashing coda, the slam and the buzz and the gull-scream.

Ring the angelus for the salmon and the swallows. Ring the bells for frogs floating in bent reeds. Ring the bells for all of us who did not save the songs. Holy Mary, mother of God, ring the bells for every sacred emptiness. Let them echo in the silence at the end of the day. Forgiveness is too much to ask. I would pray for only this: that our granddaughter would hear again the little lick of music, that grace note toward the end of a meadowlark's song.

Meadowlarks. There were meadowlarks. They sang like angels in the morning.

The populations of meadowlarks are declining, and sometimes plummeting, across the United States: for the eastern meadowlark, a 70 million decline; for the western, a decline of 60 million. The primary causes of decline are widespread pesticide use and habitat loss, as their prairies are converted to cropland and fracking pads.

(Cornell Lab Ornithology)

The Terrible Silence
of the Empty Sky

INTERGALACTIC SPACE

I HAVE NEVER HEARD ANY MESSAGE FROM BEINGS FROM OUTER space. Well, I've heard voices from the moon ("the Eagle has landed" and "a giant step for mankind [sic]"), but those don't count, because they are Earth-based voices. I heard an urgent message from the atmosphere itself when I heedlessly walked toward a boulder on a mountain pass in a thunderstorm; the air actually screamed, rising in frenzied pitch, the closer I got to the boulder. In ice-solid cold at the Arctic Circle, I heard northern lights crackle and snap. But I have never once been contacted by an intelligence from beyond our planet. Neither has any other credible listener, as far as I know, even though the government has spent dozens of years and $20 million a year listening with all its most powerful antennae, and the Paul Allen Foundation has spent around $30 million trying to make contact.

This silence is really, really odd.

It's not what one would expect, my friend Martin Fisk told me. If you calculate that there are somewhere between 100 and 400 billion stars in our galaxy, which is just one of 100 billion galaxies; and you take into account that many of those are likely to have Earth-sized planets orbiting around them in zones with habitable temperatures; and you figure that many of these planets are 12 billion years old, which gave them plenty of time to evolve intelligent life; then you would expect that some life forms have emerged on a significant percentage of them. If intelligence has survival value in at least some situations, then it will develop. So it's an excellent bet that there is intelligent life sprinkled all over the galaxy, Marty says. In fact, astrophysicist Frank Drake and his colleagues, who did the initial calculations in 1961, estimated that there are perhaps 10,000 communicating civilizations among an assumed 400 billion stars in the Milky Way galaxy.

So where is everybody? That's the question. It's such a famous question that it has a famous name, the Fermi Paradox, after the nuclear physicist who first raised the point.

Marty lifts his eyebrows and gives me a quizzical look. He's a trim man with a neat gray beard and a trickster smile, like he's one of the few people who knows that the universe has a sense of humor. He teaches astrobiology at Oregon State University and was a member of the Mars Science Laboratory team. Because he likes the same things my husband and I like (wine, weird facts, mountain trails, etc.), we talk sometimes, most often walking the bike path through the college's agricultural lands in our hometown. That's what we're doing tonight, heading over the covered bridge and out toward the pig barns.

I don't know, I answer. Why haven't we heard from them? Earth has been sending out AM radio signals for eighty years, even before the first broadcasts of *I Love Lucy*. Where are the laugh tracks from outer space?

On this late fall afternoon, when light glances off the metal barn roof and skids into the branches of the oaks, it's a fun question, a game of inferences. But I can tell you: On the way back, after dark, when Venus is the only hole in the night and the oaks scratch their fingernails against the black ceiling of the sky, the question unsettles me. I choose my steps carefully in the dark.

Either we are alone in the universe, or we are not. If we are the only creatures in the universe who can make meaning out of all of this, what a terrible, terrible silence surrounds us. And what a soul-crushing responsibility we have, not to blow the one chance the universe has to understand itself. On the other hand, if we are not the only creatures, why won't they talk to us? There are many answers to that question, and some of them are not good.

So, okay, I say to Marty. What makes you think there's anybody out there who can send a message? Let's assume that intelligent life has evolved in any number of places and forms in the universe. Maybe there are increasingly complicated algae or bacteria, let's say. But why should any of those life-forms evolve the technological mind, and the technophilia, to extend their intelligence into space, just because we did?

Interesting, Marty says. Intelligent life evolved at least twice on Earth, he points out, and who knows how many other times. On Earth, it took 3.8 billion years for human intelligence to evolve from the first evidence of life in rocks, and 500 million years for octopus intelligence to evolve from our last common ancestor. As Peter Godfrey-Smith pointed out, communicating with an octopus may be as close as we will ever get to communicating with an intelligent alien to our own.

But the point is significant. The octopus will not send out radio waves, and neither will any other intelligent marine creature. They can't have fire, so they can't have combustion, so there you go.

Yeah, I say, warming to the subject. Think of other reasons why an intelligent creature would be silent. Maybe they are hunkered down under strict blackout orders, in fear that whatever intelligence discovered them would be nasty predators. Or maybe they evolved without any curiosity at all, or evolved such tremendous egos that they assume the universe was made for them alone. Or maybe they are busy. Busy happens. Maybe they are waiting to develop intergalactic communications until they have created the perfect opera, embarrassed to be represented to the universe by something less than wonderful.

As the light fades, black-and-white cows press against the three-strand fence, eager to return to the barn. They may be frightened of night, when coyotes come out of the culverts, or maybe their udders ache with the weight of their milk. They low piteously and drool. On the other side of the pasture, the lights from the barn are yellow and cold. A radio is on in the stalls, but from that distance all we hear is noise.

Or maybe the aliens *are* answering the signals we have been sending into space for the last eighty years. But the distances are so great that their answers haven't reached us yet. Maybe we just have to be patient. And it's worth being patient, because if they have the technology to hear us and send an answer, they surely have technologies more sophisticated than ours and civilizations productive of surplus enough to support them. We could learn plenty from them and, who knows, maybe their advice could save us. Maybe just knowing they are out there, talking to us, would bring all humankind together in a defensive crouch and inspire us to clean up our act.

It's good and dark now on the path between the pastures, and we turn back toward town. A mass of sheep huddles under an oak in the field to the right, but their darkness casts no shadow and I sense rather than see them. What I see, off in the distance, is a thick column of yellow light rising from the bowl of the football stadium. It looks like a UFO that flipped upside down on landing and now directs its spotlight into the sky, rather than down to Earth. From the bowl comes the faint music of the marching band, practicing for tomorrow's game with Stanford. Here, a mile away, the John Philip Sousa of a marching band sounds brave and good, but from a distant planet, I worry that it might sound aggressive.

The possible explanation for the terrible silence that disturbs me most is the obvious one: Maybe there are plenty of technologically savvy civilizations out there among the stars, with more evolving every eon. Maybe the cosmos echoes with the sounds of drills and hammers and martial music. But maybe every single civilization, when it reaches a certain point of complication, develops the means to destroy itself, and does.

Take just one obvious example. If burning fossil fuels is a necessary condition for a technological civilization, and if burning fossil fuels is a sufficient condition to bring about its self-destruction, then of course, instead of hearing greetings from aliens, our powerful receivers will pick up only the echoes of pumpjacks, robotically pounding the dead planet and spouting oil, long after their engineers have smothered in the heat of their own creation.

But there are surely some technological civilizations that do not have fossil fuels, or do not use them and find their power some other place. Marty stops walking and considers. His voice comes out of his black silhouette against the blinking blue light of an emergency phone attached to a telephone pole.

Okay, first of all, any extractive civilization is finite, right? That just stands to reason. If your civilization requires finite resources, your civilization itself will be finite—doesn't matter what resource that is. Will all planets have fossil fuels? Not necessarily; even if a planet evolves carboniferous plants, a fungus may evolve to turn the dead plants back into carbon dioxide. Or some beings could produce their own energy; a planet where the intelligent beings are trees would be all set, transforming sun directly into life energy. The critical question is, of those alien civilizations that turned to fossil fuels, have any of them found the will and the means to turn away?

Here's what I am willing to bet, I say. Maybe there are civilizations that are spinning along just fine, having somehow learned to live humbly and within their means by celebrating simplicity, developing fine and practical arts, harvesting sunshine, and honoring the creatures that share their fecund planets. We will not hear from them, because they would have no interest in building the machines that shout so loudly that they can be heard on the far side of infinity.

But throughout that same infinity, I wager there are no civilizations of intelligent, tinkering beings who persist in believing that they are better than their planet, that it all belongs to them, that there are no natural limits to the taking, the incessant taking, that invention is the solution to every problem—including problems the aliens create by destroying the necessary conditions for their own survival. These beliefs are lethal to civilizations. We will not hear from the people who hold to this hubris, because their civilizations are dead or soon will be. This accounts for the terrible silence of the night sky.

The steers are clanking in their stanchions as we walk by the barn. I imagine their thick necks pressing forward and their long tongues reaching for the last grains in the seams of the feeder troughs. There had been barn owls in that barn, before the remodel.

We are approaching Thirty-Fifth Street, where the university's agricultural lands end and the campus begins. Now and then, the headlights of a car plow a path through the dark. What are the chances that Earth-humans will be the first creatures in the universe to figure out how, as an act of collective will, to pull back into the limits of their beautiful planet and go silent?

Not good, I say.

But Marty wants me to think about deep time. How often has developing life started over? How many chances will it have to do it again and again? An astrobiologist thinks in terms of emergent properties, the endlessly branching oak of possibilities. Consider Earth, he says. Humans may not be Earth's first rodeo. The Earth is 4.5 billion years old. How long ago would a previous Earth intelligence have to have existed for its fossil records and DNA to be completely erased? He answers his own question: If intelligent life existed during the Earth's first 500 million years, which is quite possible, all evidence would have been destroyed by subduction into Earth's molten core.

It's evidently true that in another 5 billion years, the sun will enlarge and engulf the Earth. But that gives evolution 9.5 billion years, which is a lot of time. If time is so incomprehensibly expansive, then possibility is unimaginably expansive too. Marty wants me to think about scale. What is happening around us that is too small for us to notice? What is happening around us that is too big for us to register? Maybe the pulse of life, the eons-long, universe-wide cycles of extinction and regeneration are the sound waves of the universe, and we are too small to hear. Maybe the universe is not silent, but is throbbing with the slow, slow music of mistakes and redemption that necessarily come with time. Maybe, or maybe not.

Marty turns off on his own street and heads back to the warm light where his wife will be reading. I walk on to our house, a few blocks east, where our own porch light illuminates the steps. Frank will be inside, reading the news. I sit on the step. We have rigged small lights on the front walk that direct light down to the bricks. They look like little round UFOs, investigating grout.

If there is a sadness as big as the galaxy, I feel it now. As far as I can tell, we don't need advice from the aliens. We know what we have to do to save civilization from self-destruction. We just don't know how to muster the collective will to get it done. What a crushing disappointment we must be to alien creatures on distant planets, who are listening for some signal of wisdom or hope.

Between 2018 and 2019, the number of reports of Unidentified Flying Objects (UFOs) and strange sightings in the sky jumped from 3,395 to 5,971, according to the National UFO Reporting Center. California had the most sightings, at 485.

(ABC News)

The Terrible Silence
of the Empty Sky

FOREST

SOME TIME AGO, IN A FOREST WEST OF TOWN, FORESTERS LED schoolchildren on a field trip through an Oregon squall. I came along, holding my daughter's hand. Gusts of wind blew through, tumbling the crows. Hemlock boughs rose up and down, as if they were patting alders on the head. The children jitterbugged up the trail to a towering tree, where someone had nailed a sign.

DOUGLAS-FIR

Height: 124'
Board feet volume: 1,060
No. of 8' 2 x 4s: 208
Rolls of toilet paper: 6,890

That's an impressive amount of toilet paper. But I can tell you, when the children pulled back their rain hoods to look toward the top of that tree, they didn't see toilet paper or hear 2 x 4s.

They heard the nuthatch *nert nert* and the scolding squirrel. They heard young thrushes trying to whistle as beautifully as their fathers. The wind shuffled the treetop, and the children heard this too, and falling cones tapping on salal. They heard a branch fall through the tree. When a Pacific wren sang its small song in the swordferns, they closed their eyes and listened, as they had been taught to do.

They saw raindrops on draped lichens, a pine squirrel sitting on its haunches, an upside-down nuthatch, and a rain of pine needles when the wind gusted through. One of the children spotted a turkey tail fungus, spread like Thanksgiving.

They smelled a sweetness that reminded them of Christmas.

They felt—how does a child feel in the presence of this majesty and music? A small one ran over and gave the tree a hug, and then they were all jostling for their turn, as the teacher tried to corral them back onto the trail.

But somehow, it becomes possible, through the terrible anesthesia of our time, to learn to look at a forest and see only commodities. It becomes possible to breathe sweet forest air and smell only diesel fuel and money. It becomes possible to listen to wrens and hear chainsaws. If a person develops the capacity to not see, to not smell, to not hear the ancient living forest, it becomes possible to mistake it for something far less and think nothing—nothing—of cutting down its trees, hauling them away, bulldozing what small, stunned lives remain, and spraying poison on the wreckage.

NOW, IN MY HOMETOWN, ANOTHER GROVE OF ANCIENT TREES IS gone. Fifteen mossy acres of an ancient community of towering trees, woven birdsongs and spiderwebs, traded away for $460,000 to help pay cost overruns on a new building in the College of Forestry. One of the trees was 420 years old, a sapling in 1599, when Shakespeare was sharpening a quill pen to write *Much Ado about Nothing*, and composer John Bennet published his *Madrigalls to Foure Voyces*.

Nobody would take a child to that forest now. It's too dangerous: the stickers, the smoldering piles, the poison, the precarious leaning deadheads, the soul-searing vandalism. But if the children were to go there anyway, what would they find?

If they listened, they would hear, not scolding squirrels, footfalls on moss, dripping fog. Instead, just the terrible silence of the empty sky.

If they looked up, they would see, not green branches decorated with swags of robin-flight. Instead, just the terrible emptiness of the sky.

What would they feel? Not the comfort and exhilaration of the breathing lives, but the terrible sadness of the empty sky.

We, neighbors, mourn the passing of an ancient grove that for four centuries quietly, steadfastly sheltered us and made us glad and grateful. But it's a far more crushing sadness to learn now that those awful deaths were needless. The university's forest manager blamed the cutting on a "mistake," the result of misunderstandings among staff.

So okay. I understand mistakes.

As we philosophers are wont to say, "Shit happens."

It does indeed.

But sometimes institutions deliberately create or tolerate the conditions under which shit is more likely to occur. That changes an accident into a moral failing.

This is the sort of mistake that "happens" when forest managers are unable or unwilling to escape the entrapments of dangerously inbred, outdated, and self-dealing assumptions of industrial forestry. Anyone who holds the fate of a forest in his hands has a moral obligation to truly understand the deep existential relation of forests and the future of civilization in this time and place.

THERE ARE THREE THINGS THEY, AND WE ALL, MUST NOW UN-derstand:

1. We must understand that we do not have the luxury of living in or-

dinary times, but rather the responsibility of living at a unique hinge point in history, the last time that humankind will have a choice: to continue with business-as-usual and watch the planetary systems that support our children's lives fray and fall apart, or to participate in the greatest exercise of the moral imagination the world has ever seen, turning abruptly toward a better way.

We know what we have to do to save a world for the children. First, stop burning carbon. Second, suck carbon dioxide out of the air by planting 3 billion trees, actually now 3 billion and 160 trees because of what the College of Forestry destroyed.

In this time, it is not morally possible to routinely cut forests, transmogrify them into obscenely big houses and unneeded university buildings, haul them away in diesel trucks, and burn the broken limbs in mushroom clouds of oily smoke from the hills behind town.

What is morally necessary is to plant trees, to save trees, to recover the birdsongs and the soil. We must empower forests to save our skins, on the chance they might also save our souls.

2. We must understand that we are living in a time of extinction. In this time, it is not morally possible to crush the nests of owls—the spotted owl and its pure, questioning call. It is not morally possible to bulldoze wildflowers—the lady slipper orchids and the trillium. It is disgraceful to mistake biodiversity for a cash account and trade it in to pay reckless debts.

What is morally necessary is to protect all the lives that are left and all the sinew and grace notes they bring to the world. We must transform campuses and yards and especially ancient research forests into refuges, to save the beings, large and small, that can still, possibly, escape through the crushingly narrow hourglass of our time. What is gone, is gone forever. What is left—that's what the world will be made of.

3. We must understand that a forest is not just a bunch of trees. *Forest* is the word for the process of weaving lives into an endless madrigal, weaving

green and squawks and green and chitters and night-hoots and black with red spots, creating a complex whole, where each watery warp and fledgling woof, each part, is astonishing and alive and worthy, worthy, the spar and the song, the rain, the wasp, the children—nestlings, saplings, fingerling trout.

On this planet, the glittering urgency of ongoing life is as close as we will get to the sacred, and forests are its jubilant expression. It follows that every suffering, every destruction of an old forest, is a profanity. It is a violence we cannot even begin to measure because we have only the sorriest understanding of the forest's multitude of lives.

Be prepared for anger and for grief. Be prepared also for gratitude. Be prepared with the strongest moral resolve to guarantee that no more ancient forests will be sold out from under our children. They are not ours to sell; they belong to the future of the everlasting Earth.

IN THE CLEAR-CUT PATCH WHERE THE UNIVERSITY FOREST ONCE grew, neighbors have pulled the pile of forest litter and broken limbs away from the ancient stump, to reveal its 420 concentric rings. They mark the years of forest fires, the decades of drought, the year the settlers brought goats to clear the brush, the year the pandemic swept like a dark wing over the Earth, the year the chainsaw with a forty-eight-inch blade sawed the tree down. Eventually, the neighbors will put a funeral marker on the stump. But for now, a benediction written in a child's spidery hand is tucked into a crack in the bark:

To the trees, from Lucia.
I hope the baby trees grow strong and the baby ferns grow pretty. I hope the little ravens learn to laugh. I hope the little owls learn to hoot. I hope that when children are afraid, they can go to the mossy old grandmother tree and get a hug.

The planet has about 9.9 billion acres of forests. They absorb about 25 percent of greenhouse gases emitted by human activity each year. However, many of these forests are threatened by bark beetles, drought, and deforestation. In Texas alone, in one year (2017), insects killed more than 300 million trees. About 10 million acres of forests are cut annually in South America.

(NYT interactive website on climate change and forests, Fourth National Climate Assessment)

The Terrible Silence
of the Empty Sky

SEASHORE

I.

TO THE NORTH, I COULD SEE ALL THE WAY TO THE BELL-BUOY AT the mouth of the jetty. To the south, the beach disappeared in fog. The winds were calm on the beach, although a storm had passed in the night, driving the tide to the base of the bluff. The tide left a row of sea wrack as tall as my boots and as long as the beach—roughly braided kelp where gulls stalked, squeaking and chuckling, probably hoping to find a dead crab with dangling legs or a buoy paved with blue mussels. At the bottom of the beach, low waves pushed clumps of seafoam. A flock of western sandpipers scurried along the sea edge there, probably hoping for sand fleas. Wearing a yellow raincoat, a boy wandered the beach between the high tide line and the falling tide. He carried a blue plastic bucket, and I imagined he was hoping for whatever he could find. He leaned over now and then to pick up a treasure and plop it in his bucket.

The little boy was whistling. Whenever he is content, he is whistling. The sandpipers piped. A gull threw back her head and brayed. Picking over a pile of rocks, three oystercatchers sharply peeped. It was a lively sea chantey they made, the running chirps of the pipers, the boy's rambling tune, the muttered accompaniment of the gulls, the bell-buoy chiming in. Nothing was missing from this morning: all the members of the choir had found their places and their parts, and maybe that was why the music was such a joy. I sat on a drift-log, grinning, as the voices came together to shake the air into something lovely and pure.

The fog blew up the beach toward me, and the world turned gray. There was the long gray expanse of sand, scalloped with seafoam, and the long gray expanse of bluff, scruffy with pines. I leaned forward to keep the little boy in view, but he suddenly paused and took a few steps backward. His whistling stopped. I couldn't make out what he had found, but it looked like a disheveled black form, matted in blown sand. The boy dug at it with a plastic shovel, trying to pry it into his bucket. It took him several tries; the form was large and the shovel was small, and he seemed uncertain. I went to him as he turned toward me. What he had in his bucket was the limp form of a black-and-white sea bird, a common murre. He wanted to bring it home to his grandfather, who would help him bury it in the back yard.

I had been expecting this dying. Murres have always been one of the most numerous of marine birds—maybe 13 to 24 million at their peak. They nest on sea stacks in great aggregations, laying pear-shaped eggs directly on rock ledges. Even before the nestlings are fully developed, they all take to the sea in great rivers of birds. Now everything depends on finding fish to eat. They dive, going down sometimes a hundred meters. Glistening in silver coats of bubbles, they beat their little wings and swim around after prey. Flounder. Pollock. Sculpin. Capelin. It's a high-risk hunt. Sometimes murres get caught in fishing nets and drown. Sometimes they are poisoned by oil spills and toxins that accumulate in their fat. But the biggest risk is that now, in the warming, souring ocean, they will not find the fish they need to feed their chicks and keep themselves alive.

On California beaches to the south, dead murres have washed up by the hundreds. They were starved. On beaches to the north, hundreds of dead puffins floated in rafts or washed onto shore. They also were starved. Two years ago, the murres off this Oregon coast raised no young. None.

But only last week, I saw hundreds of murres on a sea stack offshore, standing shoulder to shoulder like penguins. The noise engulfed the headland, waves smashing, murres growling a ratchety *urrrr*, every bird fussing and yakking, the crowd noise growing, growling, *g-r-r-r-r*. Now and then, a gull squealed.

Whether all those murres were refugees from hungry seas to the south, I do not know. The seabird populations are blotchy—present one year, absent the next—which makes them hard to track. But it is clear that the food webs of the ocean are no longer supporting the small fish these birds eat, and sooner or later, that will be the end of them. I had hoped it would be a long time before the great starving would reach the Oregon coast, but this might be the beginning of it, in this blue bucket.

I did not talk to the boy about this. Carefully, we emptied the bucket onto the polished log. Here was the body of the murre, feathers disheveled, head hanging. Here was a blue by-the-wind sailor, a small jellyfish with a wing to catch the wind. Here was a mudstone, shaped by the fossilized shell of an ancient butter clam. A broken Styrofoam float from a drift net. The hinged halves of a razor clam. Half a sand dollar. A rockfish spine, chalky in death.

The murre would rather be buried here, at the beach, the boy decided on second thought, in the place it knows and loves best, where its mom and dad are probably nearby, looking for him. So this is what we did. Under the terrible silence of an empty sky, we dug a hole behind the log, buried the murre in sand, and decorated the grave with the treasures the boy had found. Looking out to sea, the boy stood straight and whistled a sad song for the dead bird. The sea-surge sang a simple hymn.

II.

UNSEEN IN THE DARK CURRENTS BEYOND THE SEA STACKS, A lost gill-net drifts unmoored, a half-mile-long curtain of gray mesh hanging from Styrofoam floats. Its hem curls and straightens in tidal currents, catches on kelp, sweeps through schools of herring. Unwary in the striped light, a salmon nudges into the net and jams there, unable to swim forward or back away. It is caught by gill plates bright and round as the moon, cratered with the moon's shadowed seas. More salmon nudge into the net, flashing silver as they struggle to escape.

A common murre swims toward the net. His head slides through the mesh, but his body is too wide to pass. He flails his wings to back away. The net snags his feathers. The fibers dig into the flesh behind the wings, deeper as the murre tugs to escape. His clawed feet scrabble at the net, yanking until feathers snap and his blood pinkens the sea. There he hangs by his neck. Behind him, a flock of murres chases herring through the kelp stems. The herring flick through the net. Murres follow. Each bird wedges into the net and begins to struggle. The more they flail, the more tightly the net seizes them. The nets bulge and recoil. Silver tails and black feathers swirl.

Tasting blood, a salmon shark sways close to the net, singing his rough flank against the fibers. He snatches off a drooping head, snatches another. But then he veers and noses into the net. He catches first a tooth, then, pushing forward, catches another. The shark whips his head from side to side, savaging the net, driving the falling scales into silver swirls. He vomits black feathers and trailing intestines that sink through the currents. A gray cod snaps up the falling pieces and pushes into the net, where she finds her own death. Heavy now with the dead, the net slowly sinks until it settles, swaying on the floor of the sea.

A hagfish sucks tentatively for torn flesh. Dungeness crab move in, scuttling sideways. A small sculpin thrusts her spiked head into the red tissues and spins, tearing off flesh. The water is cloudy with sea fleas and shrimp eating the soft meat under the sodden feathers, nibbling around the bones, a cloud of eating. Spot shrimp stalk in on spidery legs, following their orange prows. Long antennae reach toward the dying and the dead. Bubbles pop from shrimps' mouths and stream toward the sun. When the banquet is finished, there is no flesh, only skeletons and wings of pinfeathers, swaying. Without the heavy flesh, the net rises again on its float.

In the tick of small teeth and the click of small claws, skeletons with feathered skirts ride the ghost net, hissing. Strips of skin swirl. Plated heads grin. The ghost net floats on great tidal currents, gathering bones.

III.

THE SMALL BOY'S GRANDFATHER, MY BELOVED FRANK, IS DYING, although you would not know it to look at him. He still stands tall and strong and broad-shouldered, ruddy as a ponderosa pine. That's what his grandfather feels like to the boy, when he hugs him around the waist—like a tree. The boy tells him about the murre—not crying, but wondering at the mystery. Together, they find a field guide and shuffle the pages to the bird. *An abundant, penguin-like bird of the cooler northern oceans, the Common Murre nests along rocky cliffs and spends its winter at sea.*

Yes, that's the one. The boy and his grandfather, seated close together on the couch, lean over the picture. But soon they are talking about penguins, and then they are talking about dinosaurs with brightly colored feathers. The boy reaches an arm around his grandfather's neck. They know things about birds and dinosaurs, the boy and his grandfather, and that's one of the ways they love each other.

It's an odd cancer, maybe a blessed sort of cancer, that has no treatment and no outward sign. It works quietly behind the scenes. That terrible silence again, like climate change, his daughter says. The grandfather and his family enjoy these beautiful, warm days—*odd weather for fall*, everyone says, *but I won't complain.* They enjoy the lack of rain, which gives them an extra month of glorious autumn leaves, and still there are roses in the hedge. And they are glad for the warmth while it lasts, knowing that something is terribly, terribly wrong, and soon—a decade, half a century—the changes will turn on the world and the underlying wrong will turn deadly. No one talks about it. No one wants to think about it. But they love the world all the more. They hold the red huckleberry a little longer on the tongue. They breathe maybe more deeply at the edge of the sea, follow skipping flocks of sandpipers with their eyes, counting. They stroke the madrone tree's skin-smooth bark, thinking, *it will come to an end, but not yet.*

Silently, unseen, the bone marrow will clog with fibers, sending out misshapen red blood cells and useless white blood cells, and the grandfather will get more and more tired and then he will die. Or maybe pneumonia will take him down. Or maybe the virus. A couple of years? Three? Maybe more. Fewer? Who knows? No one talks about it, but everyone thinks about it, and of course their love for one another grows sweeter, deeper, and there is much more touching. The grandson snuggles closer and whistles a song about a sailboat. The grandfather's grown children hold him tighter before they leave for their own homes.

There is such a thing, I am sure, as anticipatory loneliness, the loneliness that comes in advance of loss. I don't know a good name for it. A philosopher might follow Immanuel Kant and call it *Prolegomenon to Any Future Loneliness.* Turn to Latin and call it *presolitudinem.* Or let it wash over you, unnamed. Maybe this is it: The sadness that forces you to close your eyes, when your ears are filled with the fuss and growl of a colony of murres, and your mind is battered by the expectation that the sea stack will soon fall silent. Or the terror you feel when you watch wildfire creep toward a ponderosa pine forest you have loved for all your life. That terrible loneliness, when you look at his strong back, to wonder how you will live in a world stripped of what comforts you, protects you, brings you joy.

IV.

IN THE DARKNESS IN THE ELONGATED CAVERN AT THE CORE OF the thick yellow thigh bone, the marrow is a sponge of soft tissue. Liquid sloshes there, carrying stem cells like coracles sailing on a wine-dark sea. But not on the open sea; rather, they bob along through a complicated cave of galleries and tunnels, dead-ends and arcades. The stem cells are pods of potential, poised to become red or white blood cells and platelets. But there in the dark, things go badly wrong.

A platelet looks like a microscopic starfish. Thick arms stick out from a center core. With entangled arms, one platelet sticks to another, and another, the way starfish will do, coming together in congregations, sucking onto each other. Massing platelets make a blood clot or a scab. But in this thigh bone, there are too many, too many platelets, and they gum up the marrow and thicken the blood. Fibers grow like stalactites into the marrow, tangling in the tunnels.

In the thickened marrow, a stem cell divides and becomes a red blood cell. Although it should look like a red snowball, it is deformed, flattened as if half-melted. It can't carry its load of oxygen to the cells. When it divides, it creates another deformed cell, which divides, creating another. But there are never enough, and the populations of red blood cells catastrophically fall and fall.

At the same time, the stem cells are creating too many white blood cells, not round as they should be, but shaped like tears. When infection invades the blood, they will not have the strength to fight it off. They swim uselessly among the crowds of other cells, reproducing wildly.

When scarring finally spreads through the spongy tissue, hardening it, it slows the cells. Tangled with fibers and blocked in the stony couloirs and narrowing tunnels of the bone, the cells will stall there, a thick mass of snowballs, starfish, and tears.

BUT FOR NOW, THE BOY STILL HEARS HIS GRANDFATHER'S HEART-beat, as he hears the slosh of waves on the shore and the laughter of the gulls. He lives in the music of the moment, and for now, this is enough. It is more than enough. It is his comfort and his delight. This morning, he was back at the beach again with his bucket. There was an especially high tide last night, so he was excited about what he might find. He was looking for decorations for the Thanksgiving table. A small black feather. A broken butter clam. The polished hinge of an oyster. Half a sand dollar. The broken spiral of a snail. A clump of empty barnacles. All things dead and beautiful.

This is what the grandfather sees, as he stands at the head of the table with his family around him, all these treasured relics. The boy has placed them randomly on the table, as the tide would have left them on the beach. *We give thanks*, the boy says, as he has been taught. For now, for what we have been given, *we give thanks*.

About 17,000 Americans suffer from the bone marrow cancer called myelofibrosis, which is associated with toluene, a chemical found in crude oil. Toluene is used in histology research laboratories, paint thinners, and other places.

(International Agency for Research on Cancer)

Twelve Heartbreaking Sounds That Will Remain

> . . . [T]he beauty of things was born before eyes
> and sufficient to itself; the heartbreaking beauty
> will remain when there is no heart to break for it.
> —ROBINSON JEFFERS, "Credo"

IF THERE COMES A TIME WHEN HUMAN VOICES VANISH, MUSIC will still ring out from the Earth. Beethoven's symphonies are already sailing toward the stars, where they will soar forever. If there comes a time when the voices of songbirds are stilled, the music of other, maybe new, beings will bell or mumble from the mudbanks to the mountaintops. The mountains will continue to roar, and the seas will sing on the sand. Earth will evolve new forms of wild music that we can never imagine but long to hear. In the next New World Symphony, others will play the parts, but the music will remain.

1. In the silted harbor, the moan of the rusting whistle-buoy, its hydraulic cylinders transmuting the lift of a wave into a cry in the fog.

2. Hard rain, fortissimo, eroding the mudstone footprint of a girl running on the beach.

3. Shards of colored glass tinkling down the marble steps, having fallen from a soaring arch past the gentle face of a stone saint holding a bird in his hand.

4. Wind whistling under a slab of bark in a forest of sycamores grown strong on the desiccating backs of cedars. It's a sound that might once have been mistaken for a thrush.

5. A steady roar, as fire flares off methane leaking from an abandoned well, the bodies of ancient dragonflies transmogrified into fury.

6. Boulders clattering down a cliff.

7. The grunt and swoon of sex. As long as there is lust, there will be music.

8. The clap and crackle of a prairie fire, a sound like a round of applause.

9. A leggy magnolia falling in the forest, the long sigh and final thud. A toppling tree makes a sound even when there is no person there to hear it fall; the whole world shivers, and the dent in the air travels around the globe.

10. The mumbled prayer of a thunderstorm, approaching.

11. The arpeggio of a creek falling through red rocks. It once was the canyon wren who sang this song, but he learned it first from the desert freshet and this is what shall remain.

12. Wind whistling through the deer-tibia flute that children left, half finished, on the beach.

Fully a third of the gas produced around the world is burned off at well sites in a process called "flaring," producing flames visible from outer space. In one year (2012), flaring in North Dakota alone released as much CO_2 pollution as a million cars—4.5 million metric tons.

(*Scientific American, Science Daily*)

Sorrow Fired to the Strength of Stone

To: Sarah
Subject: RE: Grief

FULL MOON AGAIN TONIGHT, SHINING THROUGH THE HEDGE, throwing broken light on my bedspread. Unable to sleep, I am thinking about your message, thinking about the grief you felt in your backyard last night as you curled in a hammock under the trees, watching the moon. Was it the quiet that made you so sad? Maybe you heard the whir of distant traffic but none of the whistle-mumbles of birds fluffing, reshuffling, settling in after a dream. Was it the silence that answered the great horned owl as all night long, he softly called for his mate? Was it the emptiness of a night without fireflies, or just a few forlorn flickers by the stream? To be sad in a hammock under the full moon, that's a hard sorrow. Oh, dear Sarah.

I picture you lying there, staring up, letting your mind write all the world's griefs on the blank face of the moon—the murdered defender of a forest, an oil-soaked cormorant, a child without the strength to cry. "The grief wells up," you wrote. I picture damp moonlight in your eyes.

I agree, it's hard to know what to make of this cosmic grief. But I do know that we have to make *something* of it. Otherwise, grief will make something of us, and it may be fear or cruelty, resignation or madness.

So, Sarah, can you imagine that your grief is clay? Can you gather it into a ball and throw it against the table, throw it again and again to collapse every pocket of air that might weaken what you will make of it? This is required, this violence, to make of grief a worthy thing.

Then put the ball of clay to spinning. As it spins, reach out to touch it tentatively with a damp hand. The grief will change against the slight pressure. It will grow taller. I believe you should let this happen. I believe you can find the courage to let the grief grow under your touch.

Exert a steady pressure on the outside of the growing shape and an answering pressure on the inside. As the opening hollows the center of the ball, insert your entire hand and pull gently up. Your temptation may be to move too quickly. Patience is needed, or the grief will be shapeless and of no use. Try to stay there with it, as you transform it into a vessel. A container of any shape will do; this is not the time for perfection.

Fire, now. Take it to 1,800 degrees, until woodsmoke chokes you and the smell of glowing cinders burns your throat. There is no holding back. The grief is hardening, becoming a permanent thing—sorrow fired to the strength of stone.

Take time to let it cool.

While you wait, you can be gathering cordage for a handle. I know your courage: You will not be afraid to ask your friends for something strong enough to help hold the weight of the world's grief. Each will give you something—a child's song, a sprouting garlic bulb, a river carving a new course, wisdom from an old woman, a blue jay's raucous courage. Rub the cords together with the palm of your hand against your thigh to bind them into a single cord, adding more strands until you have sufficient length and strength. With careful knots, weave a net of cordage to hold the pot and form a carrying strap. Make the strap strong. You will carry this weight with you wherever you go.

By the time the full moon returns, or the one after that, you may be ready to go back out to your dark backyard. We can do this part together. We will collect every gift of healing that the night offers. We might start with air. Or the moisture of the night. We will carefully lift shadows into the jar; you will need them to wash their deep blue over the bright urgency of your apprehensions. We will have to work hard to get gravity into the jar, but it is worth the effort; no one will ever have a stronger or more trustworthy lover to embrace her. Look for the eye-shine of a spider in the maples and add her hungry hope. Warmth is next; we will scoop it in with the palms of our hands as if we were beckoning it into the jar. We'll try to catch a bit of time, the most healing of all Earth's gifts, and put it in the jar under the faithful tides. If you find yourself at some point smiling, hold this too. As you surmise, even together, it will take us many nights to collect Earth's gifts—there are so many, freely given—and when the sun rises, as it will, we will put that into the jar too. Tie the jar to your belt with its cord and keep it always with you.

This is important, that you carry the weight of this jar. Otherwise, Earth's healing gifts may dissipate and escape you, and you—not seeing them—might imagine yourself bereft. I am allowing myself to believe that with grief always nearby, gratitude for the world's remaining gifts may be close at hand, and maybe that will be enough to sustain you in your work— for the fireflies, for the widowed owl, for the night-singing birds, for the uncertain human future.

In 2017, the American Psychological Association coined a new term, *eco-anxiety*, "a chronic fear of environmental doom." In 2020, the Yale Center for Climate Communication found that 69 percent of Americans are "somewhat worried" about climate change and 29 percent are "very worried," an 8 percent jump over 2018. Earlier Yale polls reported that among the "very worried," 85 percent are afraid, 81 percent are sad, 79 percent are angry, 76 percent are disgusted, and 61 percent feel helpless.

(MedicalNewsToday.com,
Yale Center for Climate Communication)

3.

AWAKEN

LISTEN: NOW, IN EARLY MORNING, YELLOW-HEADED BLACKBIRDS call from the center of the marsh. They make a grating, buzzing noise like barbed wire caught on a gate in the wind. Then, *chack*, they say, *chack*. Red-winged blackbirds shrug their shoulders and cry out. *Okaleeee. Okaleee. Chack.* Mallards set their wings and land feet-first, quacking as if they were cartoons. From far away, a line of Canada geese brays as it comes into sight. The geese bank to float onto the water, honking and gabbling to beat the band, sending a shiver down their backs to settle their wings. *Chack.* It's the usual riot in Duckville on a spring morning. Cars crunch gravel as they slow to a stop. Birdwatchers roll down their windows and scan with binoculars. They wince when the water tears apart, as if in a terrible williwaw, and in a rush of splash and honk, every duck and goose and coot flaps off the water and honks away, defecating. A peregrine falcon soars, silent, over the suddenly empty marsh.

Living Like Birds

I. Living Like Birds / American Robin

NEVER IN MY MEMORY HAS A MORNING BEEN THIS QUIET NOR the air so clear. Never have spring colors been this alive—new green leaves under a storm-blown purple sky. Never has rain glistened against the hillside with that magnifying light. Two red-tailed hawks soar in a sky that is unmarred by jet contrails, unshaken by their thunder. Crows stalk down the street, pecking in the piles of pink petals washed from apple trees into windrows at the curb. Crowded together in complete disregard of social distancing, American robins wander singing through what seems like an avian garden party in the lilacs.

For my part, I am more careful in the midst of the coronavirus pandemic. I leave my house only at dawn, when I can zigzag three miles through this compact Oregon town and pass only birds. If I do spot a human being in the distance—maybe a pajama-clad man snatching the newspaper from his sidewalk or a masked woman walking a dog—I make an abrupt turn and take a different route. I make sure to be home by 8:00 a.m., which seems to be when the joggers emerge, breathing like whales. Then I stay in my house and garden, except for an occasional foray out for groceries or wine.

So I am living like a bird now. My neighbors have also become birds: Under a state-mandated lockdown, we hide in our cluttered houses behind the hedges, darting out only now and then to find something to eat, hoarding food with the cleverness of a scrub jay. We are afraid of being hungry. We are afraid of human contact, the way robins used to be, and we flutter across the street or around bushes to avoid people, knowing that we are vulnerable to every miasmic wind, that a human touch could kill us. Now and then, we sing from high or hidden places, but mostly we are quiet. We are worried we will sicken and die now, poisoned by human recklessness and political stupidity, dying as birds have done for decades.

Meanwhile, the birds' voices ring brilliantly in the quiet air, and is it my imagination that they are singing with reckless joy? Of the suffering of cities, the frightened loneliness of elders, the desperation of mothers, the panic of nurses, the bewildered hunger of children, all the suffocating old men attached to tubes—the robin knows nothing.

Sorrow is not carried on the wind; it has to be passed person to person, in close contact. So I do not expect the robin to know our sorrow, any more than he can understand why the sky is suddenly clearer or why his song carries so far. As for the sudden kick in the gut that the virus gave to the Masters of Creation, Mother Nature's Favorite Children, humanoids brought to their knees by an undead, unalive packet of genes—robins most likely can neither sympathize nor jeer.

But birds are perceptive in their own ways. The robin looks at a worm with one eye, cocks her head to inspect it with the other, then gives it a mighty peck. Like the robins, the other songbirds in my backyard have an eye on each side of their heads, so they can see into the world around them. I don't know how well they can see ahead. But I expect that they live in the necessary urgencies of the moment, without a concept of yesterday or tomorrow.

Regret for yesterday and preoccupation with tomorrow is a gift peculiar to human beings. It is a mixed blessing. Maybe there has been profit in foresight, which allows the constant calculus of means and ends, investment and outcome, human "progress." But there is also agony in foresight, to clearly perceive what lies ahead on this path we doggedly, dim-wittedly continue to choose: climate chaos, species extinction, injustice spiraling beyond human decency, and now pandemic disease. Now we are living out the betrayal of foresight, when it is possible to be destroyed by despair for the future, even on the most glorious of spring days.

So let me live like a bird in this way too: Not forever, but just for a while, let me perch on my front step and attend to the world around me, rather than the vision ahead. I will watch the robin make her nest. She drops a mouthful of twigs into the crook of the apple tree and then crouches over the growing pile, shivering her tail and what must be her elbows to compact the sticks into a bowl. Then she brings another mouthful of twigs. Then another. Spent apple blossoms fall around her and tangle in the twigs. With no thought for the future, the nestling robins and the hard pink apples will ripen together.

For this moment, I allow myself to come into the "peace of wild things." The phrase is from Kentucky poet Wendell Berry. "When despair for the world grows in me," he wrote, "and I wake in the night in fear of what my life and my children's lives may be, . . . I come into the peace of wild things, who do not tax their lives with forethought of grief." Grief may come to the birds in a spring thunderstorm, but it does not lurk in the robin's mind.

Soon, I will come back with a jolt to the global tragedy and turn, panicked, to what is required of me. But right now, I am comforted by imagining what the world looks like from the height of a soaring hawk, as the sun sweeps over the green hills, the glistening rivers, the empty streets, pushing back the night, as it has done, and as it will do, until the end of time.

II. Dying Like Birds / Barred Owl

WITHOUT CLOUD COVER, NIGHT CAME LATE TO MY GARDEN. THE moment the sky went black, the moon seemed to pop open at the top of the sky—just half a moon, and waning. It backlit the little clouds, making them look like sneezes. I was waiting for the barred owl to call. All week, he and his beloved had been hooting and clacking at dusk, catcalling in a lusty jazz duet. Maybe they had found a nest cavity in the towering Douglas-fir tree in the corner of my garden.

But I didn't hear the owls, and at the time, I didn't wonder why. I was distracted by a faint gargling from high in the night sky. It turned out to be a moonlit V of white-fronted geese, migrating north. The barking, honking brays descended from a great height, as if the clouds were laughing. I was laughing too, happy that the geese had escaped the lockdown, freely touching wings as they flapped through the night toward the pure Arctic tundra tussocks. But my happiness made me feel guilty, knowing that thousands of people were choking and calling out in the night.

"Someone will say: you care about birds. Why not worry about people?" the Trappist monk Thomas Merton wrote. He went on: "I worry about *both* birds and people. We are in the world and part of it, and we are destroying everything because we are destroying ourselves spiritually, morally, and in every way. It is all part of the same sickness."

What is he talking about, a human-caused sickness that is destroying the natural and cultural world, as it destroys our souls? It's undeniable that for decades, Americans have recognized the symptoms of something terribly wrong in our country, in the drained marshes and melting Arctic plains, the street camps where sick people sleep, the huddled immigrants, the poisoned strawberry fields and brown city air, the miles of pumpjacks and oil trains, twelve-lane highways thick with cars—the dying, all the dying, the economic investments in dying, the political collusion with dying, and the furious defense of the instruments of death. Thirty percent of all songbirds, the robins and swallows, are gone from North America. People of color are three times more likely than whites to die of the coronavirus. Those who notice the spreading soul-sickness swallow sour pills of confusion and regret.

We privileged people have not loved enough. Oh sure, we loved money. But we have not loved the poor and fragile, the threatened and endangered, the brown children and the birds. We have systematically undermined the ability of humans and other animals to get food and shelter. We have poisoned their neighborhoods and meadows. We have darkened their skies and dimmed their cities and drowned their music under the industrial din. We have allowed ourselves to think, *What is it to me if they are hungry or cold or displaced from their homes? What is it to me if their habitat is bulldozed, their means of subsistence is destroyed, and their offspring die of hunger or despair? If extinction and misery are the price to be paid for prosperity, how convenient that we can send the bill to animals and to the poor and displaced.*

But then, weakened, they die in great numbers, and we are astonished. We protest: *We loved the robins. We loved the jazz musicians and the wise elders with their rheumy eyes.* Really? *Well, maybe not as much as we loved other things.* The disastrous failure to love enough has sickened us. I believe that is true.

How long have we breathed the dust of meadowlark nests crushed in the ranchers' combines? We inhale the molecules of what the vultures leave of the truck-struck skunk at the side of the road. We breathe the exhalations of exhausted mothers working three jobs and the poisoned sweat of field workers. The treacherous fragrance of our gardens carries the ground bones of seabirds caught in fish nets. What remains of owls, ancient forests, or injured loggers in the wide cedar beams in our homes? In the course of every ordinary day, we inhale the droplets of deaths that our culture carelessly causes, and surely this is another way the sickness spreads.

My neighbor found the male owl the next morning. He was dead, splayed on his back beside the little fountain under the roses in the backyard. The owl's body was bigger than my neighbor expected, feathers fluffed, dark eyes wide open, wildly looking ahead. Rat poison ruptured his heart, most likely. Another unintended but entirely foreseeable death.

I will not be able to return to the garden tonight. I could not bear to hear the female owl crying out for her mate. But I can't stay inside, in front of the news. One hundred and thirty-one thousand people have died of the coronavirus in the United States so far. I cannot bear to imagine the crying out.

III. *Loving Like Birds / Tree Swallows*

IF I WERE A BIRD, I WOULD BE A TREE SWALLOW. THE DIET OF gnats would be worth it, to soar over water in extravagant arcs and to make love on the wing, just the lightest touch of two iridescent bodies spiraling toward the river.

I live in a college town at the confluence of two rivers that flow through a fertile valley between the Pacific Ocean and the Cascade mountain range in Oregon. The pioneers who settled here in 1845 thought they had arrived in God's garden, unaware or uncaring that it was not divine providence but devastating smallpox plagues among the Native people that left the land so beautiful and empty.

At the first sign of the coronavirus, descendants of those settlers crowded to the forest trailheads and strode through the shadows, stopping only to breathe great gulps of fresh, green air. Those who did not flee to the forest found comfort in the long, empty expanses of beach on the Pacific coast, where westerlies lift gulls and soft rain washes the air. Just the touch of that air, that's what people craved, just to feel its light touch, to consummate that great love. Of course, the search for the solace of nature overwhelmed the parking lots and trailheads. So authorities closed the beaches, then the parks, then the national forests, then the wildlife refuges, then the boat launches, leaving people in lonely misery, isolated from the sources of their consolation.

The people withdrew to their gardens, and oh, there have never been such gardens. At first it was flowers that people grew, an abundance of daffodils and tulips. But now kale and lettuce and peas grow abundantly between the spent blossoms, and never is kale so lovingly attended. In their gardens, people find life ongoing and deep gratitude for gifts that Earth continues to give, although her body is weary and her skin is flayed.

The gratitude is expressed in sharing, much as the Native Kalapuya people give thanks to Earth by sharing huckleberries and salmon. My neighbors share news of birds:

"Tonight you can watch a hundred Vaux's swifts swirl around the chimney at the end of the street and, one by one, drop in for the night."

"The chickadees are nesting in the box behind my house; come and see."

"Who can tell me if that is a mourning dove I hear in the mornings?"

"So evenly spaced on the telephone wire, are the swallows practicing an avian sort of social distancing?"

A bouquet of tulips found its way to my front step, a bag of the first nettles, a child's drawing—and who is to say which is the most nourishing of all these gifts? They all feed the same hunger, to be part of continuing life, to be part of growth and blooming, evidence of the great healing force of nature. They invite each of us to be subsumed into something far more powerful and enduring than any human grief. The gifts of nature tell us there is a persistence to life that no measure of insolence or greed can destroy.

What we are looking for, out here in the Pacific Northwest, is a love that is worthy of this sheltering world. How can we measure this love?

Love is measured in comfort and joy. The natural world holds us tight in its arms—calm as we tremble, patient as we mark the days "until this is over," strong as we weaken. When the time comes, the natural world will embrace us as we die. It will never leave us. If we are lonely, Nature strokes our hair with light winds. If, frightened in the night, we wander outside to sit on a bench in moonlight, it will come and sit beside us. If we are immobilized, having lost faith in the reliability of everything, still the Earth will carry us around the sun. If we feel abandoned, the Earth sings without ceasing—beautiful love songs in the voices of swallows and storms. This sheltering love calms me and makes me glad.

But I also believe this love is measured in grief. The more we love the robins, the more we love the frightened grandmothers, the deeper is our grief when we lose them. This is as it should be. Our grief should be magnificent and terrible, to match the magnitude of our loss.

A love that is worthy of this world is also measured in fury. By what right have human decisions drained the veins of the world, killing off fully 60 percent of its beloved small lives, plants and animals, over the past fifty years? By what right have corrupt governments withheld the information, planning, and equipment that might have saved thousands of beloved friends from the virus? It's the same sickness that destroys the solace of nature even as it creates our boundless need for it—the exercise of power and accumulation of wealth unconstrained by foresight or conscience. This cannot continue.

So a love that is worthy of this world must be measured in action. In the midst of a pandemic, we have shown that we don't need profligate consumption of fossil fuels and meat. We have shown that we can leave aside our selfish concerns and make huge changes in our personal lives—some sacrifices, some improvements—for the sake of the common good. We have shown that we can make good decisions without the leadership of idiots. We have shown that when we reduce our dreadful presence on the planet, the world rushes in to heal itself and thereby heals us, almost overnight washing the air, silencing the din, brightening the colors, growing the gardens, preventing pollution-caused illnesses, and bringing wild creatures out of their dens into the garden. This swift and astonishing resilience tells us that Earth will give us a second chance to start our culture over and get it right this time. We must seize that chance with courage and conscience.

And those little tree swallows who make love in midair: Right now, two of them, barely bigger than butterflies, are driving away a red-tailed hawk that threatens their nest. The hawk is protesting, crying *hooah hooah*. The swallows strike him with their wings and harass him with tiny claws. They pepper him with high-pitched chirps. They know without knowing that love of life is not only a comfort but a call to brash acts of courage and common cause.

No one knows yet how many people will be killed by the COVID-19 virus. By July 3, 2020, there were an estimated 11,085,543 cases worldwide, and 2,860,932 in the United States, with 526,408 deaths worldwide and 131,808 in the United States, and rising. There are more stable figures for the mortality rates of birds. In the United States, an estimated 3.7 billion birds are killed each year by cats, and another 365 to 988 million are killed by smashing into structures.

(worldometers.info/coronavirus/#countries, Audubon.org)

Sleep, Judy Francine

THE FIRST BIRD SONG IN THE MORNING WAS THE ROBIN'S TIMID question. *Verily truly?*—not really sure, but wondering, the way small children wonder, lifting their heads off the pillow to whisper, Is it morning? *Verily truly?* When I woke again, the air was awash with twitters and a sort of slurp that I knew was swallows. And then the meadowlark sang.

The night before, I had spread my sleeping bag on an eastern Oregon sagebrush flat next to Chickahominy Reservoir and settled in. The night was clear at first, warm and shiny black. But when the temperature dropped, the air thickened in an instant. I slept at the bottom of a sea, the air was that wet and heavy. Morning light didn't splash into the eastern sky. It grew out of the moisture, every droplet twinkling, the very air brightening, and that's when the birds began to sing.

The song of the meadowlark seemed happy. Why wouldn't he have been glad? Maybe he was singing from a bone-deep confidence that when morning warmth landed at the far edge of the flat, the mist would rise from the rabbitbrush. Insects would rise too, ascending into the sound of their rustling and the distant-surf roar of trucks beginning to move on Highway 20.

It was the promise of a day of bustling light and flies, a good day for a meadowlark.

How did the meadowlark's song sound? Like a flute underwater. Like light on a riffle. Like dampness on a yellow squash. A pure line of song, softened somehow, and slippery. My bird book said the words to his call were *sleep loo lidi lidijuvi*, but not this meadowlark. This meadowlark clearly sang, *Sleep, Judy Francine*. On one side of me, *Sleep, Judy Francine*. Then silence and a flutter. Then on the other side, *Sleep, Judy Francine*, as if the meadowlark were searching for her in the sage.

WHO IS JUDY FRANCINE, A PERSON SO BLESSED THAT MEADOW-larks call her name? I googled "Judy Francine" when I got home and was sad when I learned that someone with that name had died. *Judy Francine S_____, 82, died Saturday at a long-term care facility in Alabama.* Given the obituary's long list of family, I imagined she had been happy. A husband of fifty years, two children, many grandchildren, and *numerous nieces, nephews, extended family, and friends*. Surely the children often visited her in the care facility, bringing bouquets of chicory and old-field asters, the last gifts from fields that late in the fall.

How would Judy Francine have felt, if she had heard a western meadowlark sing her such a sweet song? In the thick sage-scented morning when the meadowlark sang, maybe she would have felt she was in the presence of angels. Maybe she believed in angels or maybe not; either way, that blessed. That protected and confident of the future. That suffused with comfort and light. A bird song could do that, could fill a heart with comfort.

Although the meadowlarks left the sage flats in late autumn when frost lit the thistles, they always returned in early spring, when buffalo grass began to green up. In the absence of the meadowlarks, geese began to call, shotguns rang against the hills, trains sang out, great horned owls clacked and moaned, and as spring approached again, the coots began to *krupp*.

And Judy Francine, lying by an open window in Alabama? Surely she listened as the seasonal music unfolded, anticipating the returns and departures, the melodies and rests, the growing and ebbing of the lyrical and strangely beautiful world. There would have been cicadas in the summer in Alabama. Autumn leaves would have clattered when gusts of wind blew through. As she watched the dogwoods turn scarlet and leaves blow off the golden hickories late that October, Judy Francine might have heard the lark sparrows' twitter and trill.

The autumn day she died, the sparrows would have been gathering their own strength for migration, along with ovenbirds, redstarts, and bay-breasted warblers. It would have been a crowded day in the singing sky, the day her spirit rose. Did you see this, Judy Francine? Did you see the pirouetting flocks of tiny birds, weaving and folding, specks of light swirling into cyclones and bursting into chrysanthemums, spreading into chords, appearing and disappearing as if they had no substance at all?

Maybe you were comforted by the assurance that birds would go and birds would return, that storms would come in season and storms would blow back to sea again, that lilies would bloom, grow hard with seed, and bloom once more. The music of the world was a repeating promise, a promise that harmony would be restored again and again in chords so complicated and beautiful that we could love them, even if we could not fully understand the genius of their pattern.

Birds came and went so reliably that scientists made charts as precise as orchestral scores to predict their travels. In Oregon, we pruned the roses on Washington's Birthday, a week before the first ice storms were likely to encase their stems in silver. The hummingbirds returned when the blueberries bloomed. Violet-green swallow eggs hatched when mayflies emerged. Wild strawberries ripened when the Marys River settled back in its bed after the spring floods. Salmonberries ripened when a full moon brought the herring to beaches on the coast. Chinook salmon returned when the water in the coastal creeks shone black as obsidian and alder leaves turned yellow. Sea-run cutthroat trout followed them in. The chanterelle mushrooms popped out of the ground just then, as if they knew when we needed butter-soaked mushrooms on the trout fillets. That was the time to harvest apples.

The pleasure was in the successful prediction. Yes, it won't be long until huge flocks of Vaux's swifts will vortex down the chimney of the food science building on campus. With our children, we used to lie on our backs in the parking lot at dusk to watch the spinning flock. The spiral leaned over the chimney and spilled a hundred birds into its maw, and then righted itself and spun away, only to return, and return, until all the birds had been poured in for the night and it was time for the children also to go to bed. It was fun, yes, but reassuring too. When swifts kept their swirling appointment with the chimney, all was right with the world.

Because we had well-founded faith in the regularity of the comings and goings, we loved surprises—an unexpectedly warm day in February that caught even the azaleas off guard, a thunderstorm in a Pacific valley that never got thunder, an early red-tailed hawk in the snow. They were the exceptions that proved the rules, the aberrations against which we measured the norms. It was a childlike glee to see even the world breaking the rules, which meant that there *were* rules, there was a way the world worked that we could see and to some extent understand. And so, once every two years or so, we stood at the windows as lightning bedazzled the night and scorched the transformers, and we called our friends to marvel. In the morning we reset the clocks, sure that they would keep the same time as before the storm.

Eight in the morning on March 15. That's when the vultures returned to Hinckley, Ohio, every single year of my childhood. The Lion's Club in Hinckley always held a pancake breakfast to celebrate their return. So we piled into the car—mother, father, three little girls, binoculars, thermos of coffee—and drove an hour, which was a long time when I was young. In the park at Hinckley, we ran first to see the vulture who lived in a chicken-wire cage that was carted out each spring. He had a lumpy bald head, a scruffy neck stuck with pin feathers, and bottomless black nostrils.

Grownups stood with their feet spread and their heads back, focusing binoculars on the sky. We followed their gaze, and there they were. Big black birds with their wings outstretched, floating casually in the white sky, fingering the winds with weathervane feathers. Vultures have no voices themselves, but as we watched them, we pretended that their wings conducted the choir that sang the new season into being.

My neighbor conducts the symphony orchestra in my hometown. When a concert approaches, I can see him in his living room, rehearsing how he will conduct. The score is spread out before him, the music from his amplifier surrounds him, but otherwise the room is empty. With his baton in his right hand, he conducts the invisible orchestra, sweeping up the violins with a majestic gesture, cuing the oboes with a jab at space, shifting the tempo with the slightest movement of his fingers, and when the music comes to a climax, leaning down to scoop up the sound with both hands, as if he were shoveling snow. I love to watch him.

I love to watch his rehearsals too. Once, when he cued the oboes and nothing happened, he collapsed over the music stand, muttering "oboes, oboes, oboes," because of course if the music needs oboes, it really needs them, and there is no substitute. So, "Start at seven. One and two and . . ." and this time, the oboes came in when they should have, and the music was entire.

And in a few weeks, there he would be, in a black tie and tails, striding across the stage to eager applause. He would step onto the podium, raise his baton, and when he dropped it again, there were the floating, fluting voices of the violins. Then the cellos joined in, then the oboes and French horns, all the voices in constant movement, skyward and swirling, apart and together again. The growing patterns created heart-swelling tensions that fell back into harmony, again and again, the spiraling music. The joy I felt lay in the faith that the dissonant chords would resolve into a sort of musical redemption. Surely they would. The violins might fall silent, but they would return in a burst of beauty; the cellos might climb into discordance, but they would settle finally into a place of peace. *Verily truly*, morning comes after night.

FOR SOME TIME EARLY THAT MORNING, I FOLLOWED THE MEAD-owlark through the sagebrush flats beside the reservoir. At first, he sang *Sleep, Judy Francine* from the fencepost beside a pasture. But by the time the sun finally flared over the distant volcanic peaks and spiked into the steaming field, the meadowlark must have been hungry. He set about flittering, sailing, flittering, sailing, dropping into the bunchgrasses to capture grasshoppers, and to grub around in the dirt. I did not hear him sing again that day. What I heard were the alarm calls of Steller's jays, the flocking calls of crows, and the whispered whistle of one white-crowned sparrow, who was probably hormonally confused by the length of the day, which was an exact match to the song-time of the spring. And so it was: the music of the day became the music of the years, the years.

The years: the meadowlarks grew quiet and the frog chorus began. The frogs blended their voices with the great horned owls, and the coyotes joined in—a great, rising chorus, fading again, and there were the mixed flocks of small birds, but then the snipe were back, winnowing, and the swallows were back, slurping, and the first meadowlark called, and Earth sailed in its grand orbit around the sun, which swept a long arm over the planet, cuing first the cellos, then the trumpets and the sandhill cranes.

But, dear Judy Francine, things are changing, and I am losing my faith in repeated patterns. The weather comes now and goes, and who can make sense of it? Cold places are warm, and warm places are hot. Water warms and trout disappear. Owls disappear and no oak seedlings survive. Viruses don't retreat in the summer, as they used to do. The rains don't come, and from the crowns of dead pines, ravens call down furious damnation.

Some beings cue to the length of the day, some to the warmth of the morning, and these used to coincide. But they don't anymore, not exactly, and this is what dismays me. You can't make a small change in the time that the oboes enter and expect the music to work at all. You can't make a small change in the pitch of the flutes and expect harmony. You can't make a small change in the temperature that cues insects to emerge, and still expect meadowlarks to sing.

It's all one thing, this opus.

That's what music is, the precarious, inseparable relationship of sounds, moving through time. That's what a life is, a temporary harmony. That's what a sagebrush prairie is, with all its parts in their places, singing their songs. I know that all beauty flies apart some time or another, that everything flies apart. But music is trying to hold it together. Prairies are trying to hold it together. Dear god, we are all trying to hold it together.

Sleep, Judy Francine, as the music of the world whispers your name. Slip into your dreams. Dream of redemption, when all the parts, which have been scattered, come into right relation again. We yearn to be called back in. Everything yearns to be called back into a beautiful, meaningful whole—the meadowlarks on the barbed wire, the frogs into their chorus, the cicadas into their pulsing choir, the dancing insects into their light, the reservoir into its banks, the aging women into the old fields where purple asters bloom.

In one of the first studies of the relation between migration timing and climate change, a Colorado State University scientist found that migrating birds were flying north earlier in the spring, with the greatest changes in migration correlating with the greatest warming. Other studies showed earlier migrations among birds in Alaska (32 percent), Maine (38 percent), and South Carolina (33 percent).

(Nature Climate Change, *International Journal of Zoology*)

Alarm Calls

WE'VE BEEN WATCHING CHICKADEES, LISTENING CLOSELY AFTER we read a study describing how sophisticated their warning calls are. A little high-pitched *seet* call warns of an aerial predator, an owl or a kestrel on the move. Then the chickadees freeze in place. But when a predator is stationary, chickadees give out the *chicka dee dee dee* call. The more threatening the predator, the more *dees* get added on to the end of the call. One researcher heard a chickadee add twenty-three *dees* to this call when a pygmy owl perched nearby. The call not only raises the alarm but also rouses other birds, even other species, to mob the predator, driving her away or at least preventing her from making a sneak attack.

On the one hand, this is delightful evidence of how time and evolution can create not only the ability to warn others in very particular ways but also the impulse to put oneself at risk to perhaps save others. I like to think that no chickadee can bear to quietly sneak off and save himself, that saving others is built into his very DNA. The call bursts from his bill, unbidden, *dee dee dee dee dee*.

But I am abashed when I remember all the times I have stood in the snow with my hand out, holding sunflower seeds. The little claws resting so lightly on my hand, the hard knock of the tiny beak, the swarming birds all happily calling *chicka dee dee dee*—behaviors that I took for trust—these delighted me and made me laugh aloud. I feel kind of crummy to know that the birds actually were warning each other about the big, blue-hatted predator, even as they took my seeds.

Never mind. The animal with the cutest warning call is the pika. This is a little rabbit-like stuffie with round ears, about the size of my closed hand. Pikas live in broken rocks, the talus slopes, on high mountains. All summer, they gather stems of grasses and flowers, storing them in hay piles in the hollows between the rocks. This means that you can sometimes see pikas with flower stems crosswise in their mouths, like the rose stem that Carmen holds between her red lips in Bizet's opera. The pikas sit on lichens, scanning the sky for hawks and the rocks for weasels.

The pika's alarm call sounds exactly like the *eek* you get when you squeeze a rubber ducky bathtub toy. The violent exhalation jerks the body backward and flaps those round ears. One researcher says this protects their hearing from their own squeaks. At any rate, the only way you will hear this is to venture above tree line to the broken rocks and snowfields. The pikas will squeak as you approach and then scamper into the rock tunnels. But if you sit quietly, they will come back out and recommence cutting their hay crop of bluebells and forget-me-nots. If you twitch then, you will have the great gift of hearing the muffled squeak of a pika trying to talk with its mouth full of flowers.

When our son was in middle school, we backpacked into the Beartooth Mountains in Montana, pitching our tents at the base of a rockslide. I don't remember what loneliness was bothering him, but something sent our son scrambling up rocks to a high ledge, where he leaned his back against the mountain and lingered, watching the sun set, purple and gold. When he finally came down again, his face was tearstained, and he was grinning. Probably mistaking him for a part of the mountain, a pika had climbed onto the toe of his hiking boot and squatted there, peering around. When one of us moved in camp, our son could feel the pika shudder and hear him squeak. He and the pika stayed there for a long time, still as rocks. What more could a young man want, what could more surely assuage his loneliness and move him to grateful tears, than to be so fully part of a mountain, so still and solid, so trustworthy, so blended into the fraying rays of the sun, calmed by a peace a pika could feel with its paws? He has not felt this way since, our son told us, except once when he was sleeping on the beach on the Great Barrier Reef and woke to find tiny, new-hatched sea turtles parading across his chest, a hillock to climb on their way to the shining night sea.

Pikas truly are creatures of the cold mountaintops. Temperatures above 77°F can kill them. They need the snowfields to cool them in the summer and to warm them with a blanket of snow over the rockfalls in the winter. So as the planet overheats and warm weather climbs the mountains, they are retreating farther and farther up the slopes. At some point, warming temperatures will push them to the bare rocks at the top of the mountain peaks, where they will crouch, squeaking their final warnings.

Of all the ways that animals can warn humans of danger, one of the most traditional ways is by dying. Everybody knows about the canary in the coal mine. Built to fly into the rarefied atmosphere, birds take in big gulps of air, so they are affected by noxious gases or carbon monoxide before miners are. The birds may die when the tunnels burp poison, but their death gives the miners time to escape. How fond the smudged miners must have been of the bright singing birds in their little cages. As the miners worked, they whistled to the birds and gave them little treats, I am told, and when the miners emerged at the end of the day, carrying the little bird into the light of the setting sun, how glad they must have been that the bird had lived another day in the dark, damp heat. By the mid-1980s, electronic "noses" had replaced the canaries, so the practice survives only as a metaphor. The miners, if they whistle, must whistle to themselves as they dig into the black fossilized trees.

It is this cross-species communication of alarm that most moves me. Lots of creatures communicate across species. Chipmunks can understand the warning called out by the tufted titmouse, and various species of monkeys understand each other's alarms, distinguishing the leopard-alarm from the raptor-alarm. Nuthatches understand chickadee alarm calls. Crows, jays, chickadees, terns, and blackbirds will respond to mobbing calls of other species and join mixed mobs to drive away a predator, squawking, divebombing, and defecating on the poor frustrated creature. Members of the human species can understand the death of a canary as a warning of the danger of their own impending death. Human ears can understand the silence of lower-elevation pikas as an announcement of dangerously rising temperatures and local extinctions. Can humans hear the slow silencing of the planet's wild music and understand that it is a universal expression of alarm?

In a small way, we have all been warned through the cross-species message of silence. Several years ago, Frank and I were sitting at the edge of a meadow with our backs to a thicket of mistletoe-tangled oaks and second-growth Douglas-firs. We were enjoying the spring sunshine and the rich smell that we guessed was coming from liverworts and lichens—a fecund northwestern forest funk. Small birds were in constant motion, flitting around, challenging each other, calling out, hopping onto slender branches that threw them back into the air. An altogether jolly scene, and the chatter made us glad too.

So it took us a couple of beats to realize that the forest had gone silent. *Suddenly* silent, as if someone had flipped off the switch. *Completely* silent, as if the whole system had shorted out. We fell silent ourselves, glancing over our shoulders, understanding that something was very, very wrong. We didn't know what it was, and we never found out. A weasel, a hawk, a red fox—any of these had become death itself, slinking soft-footed through the forest, making no sound. Oh, we listened but heard nothing. Eventually, a chickadee gave the all-clear—I guess that's what it was—and the forest exhaled with song. I found I also had been holding my breath.

Now, years later, I am not sure I know how to read the increasingly complex silence of a forest. There are 30 percent fewer songbirds in the forests now than there were fifty years ago, so that explains some of the quiet;

we are losing the robins, the red-winged blackbirds. Frogs have declined 38 percent. In Europe, 30 percent of the cricket and grasshopper species are threatened with extinction. Forests are more fragmented by roads and trails, so some of the quiet must be created by the encroachment of noisy machines. The meadow where we once rested has been plowed for corn, so agricultural poisons have reduced the birds' food supply, insects and seeds.

All of these account for the quiet, but I'm also aware that my own presence surely silences the birds. A sphere of quelled voices surrounds me wherever I walk, the way crowds quiet when an ambulance passes by. The silence makes me afraid. When the birds fall silent, that is the loudest alarm call of all.

Growing up in Ohio before sophisticated weather forecasts, my sisters and I were taught that when the sky turned green and the birds stopped singing, it was time to run for our lives to the storm shelter in the basement. It may or may not be a folktale that birds get eerily quiet when a storm is coming, part of the calm before the storm. But we believed it, and we listened hard, and in so many ways, we were right to be frightened in that silent green light.

Coal mining is declining faster than canaries. The International Union for the Conservation of Nature classifies canaries as a species of "least concern." However, coal production in the United States is down 40 percent from its peak in 2006. The number of people employed in coal-related jobs is the lowest in history. In 2016, 25 percent of coal production was in bankruptcy.

(International Union for the Conservation of Nature, Brookings Institute, S&P Global)

Another Marshland Elegy

WE'RE DRIVING ALONG A DIKE ROAD ACROSS THE DRIED-UP BED of Malheur Lake, Frank and I. The lake bed stretches, dusty and cow-pocked, for ten miles to the east and at least that many to the west. We're laughing, because that stretch of dike reminds us of a story we've told for at least thirty years.

That long ago, our friends were driving their two young sons on a field trip through the Malheur. The lake had water in it then, so thickets were heavy along the ditches and crazy with birds. The details are lost in the past, but we tell the story this way:

"Look. It's a mountain bluebird right there on the fence," a parent said. "Oh, it's gone."

"Look! An avocet. My god, see that rusty neck, right there on the mud flat. Oops, it flew."

"Is that a badger? Right there beside that ho . . . Oops, it's gone."

Of course, the children saw none of these, although they jumped to the window at every alert, throwing off seatbelts and slapping from one side of the car to the other, searching the unspooling landscape with narrowing eyes.

"My god, a flock of sandhill cranes," a parent called out. "Can you hear them bugle?"

"Where?"

The car lurched to a halt and everyone hopped out, ears tuned to the sky, but the cranes had dropped into a distant field.

At that, the older boy pounded a fist on his thigh, and here comes the punchline: "Okay. From now on, don't show us anything unless it's *dead*."

Back then, the Malheur Lake wetlands were a stopping point for migrating sandhill cranes and flocks of snow geese and tundra swans that filled the fields. They whooped and clattered like victorious football fans. It had been a reliable stopping point for birdwatchers too. They flocked to see the spring migrations in such numbers that TV crews came out from Portland to film the crowds. Our family was often among the birders.

One wet year, we drove the dike through lake water that sloshed over the road. For our children, the highlights were a drowned cow—bloated, with all four feet in the air—and a coyote leaping after field mice that were flooded out of their burrows. We tried to jump a ditch to get close to a porcupine that had climbed a willow to safety. The kids made it across, then turned to watch me jump onto a mudbank and slide in, softly swearing. Water birds were so thick they bumped shoulders on the water, hissing and bleating.

Over the lake, lines of Canada geese, wigeons, gadwalls, and willets approached on intersecting flight paths that would have made even the most experienced flight controller blanch. What birds didn't land, chortling, on another bird's back landed square on their own reflections. Coots *krrick*-scraped like frogs, mallards quacked like Donald Duck, geese brayed like donkeys—a deafening ruckus, like an ambulance driving through a circus band. But noise that was merely deafening became earsplitting when a bald eagle swooped over a great covey of coots in a cove.

The coots stampeded across the surface of the water, paddling with wings and feet, raising a wild wake with each running step, tearing the water to shreds. In a black mass, the flock ran right up into the air and tilted to starboard. The eagle, maybe confused, flapped over the flock and just kept on flapping until he was out of sight. The coots skidded back onto the water, crying bloody murder.

Today, as I say, the lake bed is almost completely dry. Where there had been pondweed and water are now only rabbitbrush and dried grass. Here is what Frank and I see along thirty miles of dike:

One coyote's hind leg.

One covey of quail.

About three hundred Angus steers, chewing white dust.

This is just another year in a drought that has settled over the high desert in southeast Oregon. Not much snow in winter, only a trace of rain in summer. Winds tumble a magpie and whip up dust devils in the road.

TO THE WEST OF THE MALHEUR PLAYA, IN A BASIN BELOW A HIGH rim, is what remains of Lake Abert. Since the Pleistocene, it had been an extensive, shallow lake. Because a river fed it, but no rivers emptied it, the water evaporated in place, creating a saline lake trembling with alkali flies and brine shrimp. For millennia, hundreds of thousands of shore birds stopped to rest and feed at the lake on their northern spring migrations.

I have seen that congregation, though it's been many years. As I sat on a rock at the edge of the lake, cold in my raincoat, a wave of dark birds materialized out of the sky and tipped toward the water. Another wave built behind it, and another beyond. I could have imagined the sky was raining shorebirds from the advancing edge of every cloud. They came and they came, falling in a shower of chirps. Avocets. Dowitchers. Western sandpipers. Phalaropes. Black-necked stilts. A birdwatcher once recorded fifteen thousand phalaropes in a single flock. They tiptoed along the shore, dipping long beaks into the alkali mud, or rested on the water, gorging on fleas and shrimp.

The trouble is that ranchers are pumping irrigation water from the Chewaucan River that feeds the lake. There is no effective limit to what they can take, because no one monitors their pumps. So now, by the time the river gets to the lake, there's not much water left. Unreplenished by river flow or snowmelt, the lake is drying up. As it dries, it concentrates salt and minerals. In these increasingly dry years, the lake is 25 percent salt, too saline even for the brine shrimp and alkali flies that had fed the birds. Now, from the rim, the lake looks like an astonished eye—a wide circle of junipers, broad white alkali plains, and a small circle of blue at the center. I don't know where the birds are stopping on their migrations now, but they do not stop here. How can they survive the flight north to nesting grounds with no place to rest and feed? The last time we visited Lake Abert, we watched for an hour and spotted one avocet and twenty-seven coots.

Even so, I could scarcely fathom it when I first read that the worldwide population of migrating shorebirds has declined by 51 percent in the time since our children were born. It's hard to believe. Our language doesn't even have a word for it. The populations aren't *decimated*; that's way too pale a word. Roman generals killed ten of every hundred soldiers to punish a mutinous squadron. That's decimation. But fifty of every hundred shorebirds are gone. What is that? Goddamn *pentamation*? Unimaginable. Unspeakable.

I remember giving a report in third grade about the disappearance of passenger pigeons. I told the little kids what my biologist father had told me: Once, there were so many birds that their passing darkened the sky for days, and when a flock landed in an oak, their weight brought the branches crashing to the ground. Farmers could hear flocks coming for miles. And then the birds were gone, I told the third graders. Hunters killed them, stuffed them in barrels, and loaded them on railroad cars for restaurants in the cities. The last passenger pigeon, Martha, died in the Cincinnati zoo in 1914, after suffering an apoplectic stroke. Now Martha is stuffed with cotton, lying flat on her back in a museum drawer. Her eyeballs are gone, and you can see the cotton through the sockets.

My teacher let me finish the report, and then she told the students it wasn't really true, what I said. There could never have been that many birds. When I reported my humiliation to my father, he was so angry he marched to school and confronted the teacher. "Believe it! Believe it. There were that many. Three to five billion! Now there are none."

And now my father is dead and the migrating shorebirds are on their way to vanishing. How can this be, all this dying? I imagine avocets and coots dropping like sandbags on the farmers' fields. But that's not always how birds die, I guess. There are other ways.

Oregon author Kim Stafford has written a poem, "How Birds Die." He starts out with a slow countdown. *Get caught by a kitty cat: 2.4 billion. Hit a window: 600 million. Hit by a car: 214 million.*

But soon he accelerates:

Lose your acre of breeding ground, and so circle the parking lot that was a marsh. Circle and circle, cry and cry.

At the end:

Be the last one of your kind, singing and singing.

The hopelessness of the last bird of any kind is hard to absorb. But imagine hearing its last song. Some of the most haunting, heartbreaking music ever written evokes the tragedy of a useless, pathological death. Mozart's Don Giovanni dies in a storm of thunderous D-minor chords. As Madame Butterfly stabs herself, Puccini's cellos and double basses drone a death march. The score of Strauss's *Elektra* is, as one critic wrote, "the color of blood." The lethal music floods through the concert hall; the audience drowns in it.

And birds? What is the sound of a bird dying? Frank told me that a swan dies with a small sigh. I heard a scrub jay die squawking between a cat's pin teeth. Did the last passenger pigeon coo herself to sleep? I don't know, but I am quite sure that the most tragic sound is silence. That's the sound that Aldo Leopold, a conservation biologist, evoked in *Marshland Elegy* to mourn the death of marsh birds.

> Someday . . . the last crane will trumpet his farewell and spiral skyward from the great marsh. High out of the clouds will fall the sound of hunting horns, the baying of the phantom pack, the tinkle of little bells, and then a silence never to be broken—unless perchance in some far pasture of the Milky Way.

AS SOON AS THE SUN DROPS BELOW THE RIMROCK IN OUR MAL-heur campsite, the temperature dives. We heat chili to a rolling boil, but by the time we lift the spoons to our mouths, the chili is cold again. When the stars come out—Venus on the horizon and then Electra and her six sisters, Cygnus the swan—we climb into our sleeping bags. This is not easy. First, we zip together two rectangular sleeping bags to make a puffy envelope. Into this we push our mummy bags, rated to twenty below. In long underwear and wool caps, we insert ourselves and cinch everything shut.

The night breeze comes up, carrying the smell of juniper. Much later, we hear the caterwauling of what I tell my husband must be the three-legged coyote and his mournful band. "Doubtful," he says, and gropes through piles of bedding to hold my hand. When the narrowest crescent of a moon clears the rimrock toward morning, an owl calls. "Do you hear that?" I whisper to my husband, and he does. He is always awake when the owls call. Maybe he just pretends to sleep. Eleven degrees by morning.

The drive home from the marsh is a straight shot across the high desert plains, a narrow blacktop road that arrows a hundred and fifty miles from Burns to Bend. Along we roll between what seems like an infinity of barbed-wire fences, past sagebrush flats, acres of cheatgrass, and skeletons of poisoned juniper trees. It's hard to know if the skim of white in the ditches is snow or alkali salts.

There is exactly one place to stop for gas and lunch along this entire stretch of high desert. The townsite used to be a stage stop, and maybe there were buildings here at one time, but now there is one gas pump and a café. We start thinking about the stop about fifty miles out. A good venison hash can be found there, if you don't mind the glass eyes of a dead deer staring over your shoulder from its mount on the wall. Or the stretched skin of a rattlesnake, mounted flat on a board. That rattlesnake must have been eight feet long without its head. I don't know who mounted it, but I bet it was a job. That last time I saw somebody trying to skin a rattlesnake, the headless snake struck at the knife.

The owner is happy to pump a tank of gas. That's a really good thing, because our tank is always getting low about that place in the road, after we've explored the dike roads and mountainsides for a couple of days. We watch the needle wobble toward E, estimating the distance to the station. I don't know what we would do if we found it closed. Sit there, I guess, with the hood raised to signal "help." Then eat the last of the chili and share the last beer. Then probably die silently in our sleep, our legs glazed with salt from wind across the alkali plain. I don't know. That's a joke.

Of course we wouldn't die. Because somebody would come by to help us—to feed us and give us fuel. The state has a phalanx of troopers out on the roads to help stranded travelers. Truckers too. They swerve to a stop in a cloud of country-western music and jump down to help, not because they are bored, but because they are kind. And the cowboys in white ranch trucks? They've known since they were toddlers that you never pass by some-body who needs help, out there on the empty road with the wind rising. That's what it means to be a rancher.

So why can't we find a way to save a waystation where the shorebirds can rest and feed and set off again, safe and strong? What else does it mean to be a human being? Farmers don't have to raise alfalfa in the desert; there are smarter crops that might save the water for the birds. Politicians don't have to weaken the Migratory Bird Act; there are better ways for oil companies to dispose of sludge than to hold it in ponds that attract birds. Developers don't have to drain the marshes; better to protect the marshes and charge more for the apartments with a view of a singing pond. Hunters don't have to hunt wild birds in the federal refuges; better to fund refuges with a tax on binoculars than hunting licenses. Humans have evolved to be so smart. Pity we haven't evolved to stop and goddamn think.

After Frank and I gas up and get going again, the sun is to the south, throwing shadows from every fencepost. Where an irrigation ditch approaches the road, light catches in the fluff breaking from cattails. Mounds of rabbitbrush are in seed too, and the fur of them holds the sun. But telephone lines sag empty, mile after mile, where we might have expected light to glow on a hawk's ruddy tail. No goldfinches gleam in fields of thistles. There are swarms of midges but no swallows. My binoculars sit on my lap. I stare straight ahead. No birds "spiral skyward from a great marsh." The only sound is the whir of our tires. If we had children with us, there would be nothing dead to show them.

Reliable numbers for the decline of migratory shorebird populations are hard to get, ranging from 37 percent to 78 percent. But the best aggregate claims a 51 percent decline since 1970. Pectoral sandpipers, 50 percent decline. Godwits, 70 percent decline. Ruddy turnstone, 80 percent decline. This devastating accounting indicates, says the Cornell Ornithological Labs, that the "global network of aquatic systems is fraying."

(*New York Times*, Cornell Labs)

Late at Night, Listening

ONE NIGHT AFTER SUPPER, MY DAUGHTER AND I STRAPPED ON headlamps, pulled on high boots, stuffed an extra flashlight in each pocket, and stepped out of the yellow circle of lantern light into the dark, heading to the cove. It was our third night camped on a tiny island in the Misty Fjords, scarcely an acre of jumbled rocks under a carpet of moss and a canopy of hemlock trees. Our tent only just fit on the one flat patch of duff, wedged behind a log high enough to serve as a sort of table that held our propane lantern and stove. The rest of the island was wild and inaccessible, in an archipelago of wild islands where no other boat found its way.

It was the night of the lowest tide of the month. On a tide that low, the mossy edge of the island hangs precariously eighteen feet over the water on a pedestal of rocks. Slicked with algae, concealed sometimes by sea lettuce or shiny sheets of winged kelp, the rocks are treacherous enough in daylight. In the dark, we lowered ourselves crablike, wedging our boots into cracks, our headlamps shooting wildly into the spruce across the cove.

At the edge of the water, we followed narrow beams of light through the strange land that appears and disappears with the phases of the moon, slowly rising from black water and sinking away by morning. We moved carefully, testing each footstep, staying so close together we bumped shoulders. Hermit crabs scurried out of the way. Under an overhanging rock, our headlamps spotlighted a colony of orange-striped globules dripping mucus and algae in threads to the water. Pink-striped anemones hung from thick stalks, heavy under their own weight. They shuddered when we touched them, and so did we. Orange worms hung from curling tubes, their own weight lengthening them into glistening threads. Wherever we looked, flat slabs of kelp shone in lamplight. Chitons, polished like leather, stuck fast to rock. There were acorn barnacles as big as fists, barnacles on the barnacles, and leaf barnacles with fleshy necks.

We looked around for a place to sit—not easy in the intertidal zone. Sit on barnacles, and they'll prick holes in your rain pants. Sit on kelp, and you're likely to skid off into the drink. We found some patches of rock wrack for an uneasy perch and turned off our lights.

Sea-smell rose around us, salty and thick. In the dark, we listened intently. At first, all we heard was the sea itself, the soft inhale, the asthmatic, gurgling exhale. Then gradually, spaces between the rocks began to tick and pop. Seaweed squeaked. Scritching moved in around us, tiny claws and bubbling jaws behind us, to our sides. There was a constant *plop plop* as saltwater dripped off invisible globules and bobules and tentacles and god knows what else. The loudest sound was a steady tick by my right hand, but then something big breached the water and slapped back down. I heard Erin suck in her breath. Water lapped against the shore. There was a flurry of scritching and a long sigh.

I shifted on my slippery perch, startling the whole world into silence. Erin switched on her headlamp and scanned the water.

"Eyes," she announced. "Dozens of eyes."

Startled, I stared into the water. Nothing but darkness.

"No, hundreds," she said, swinging her headlamp beam across the bay. "Hundreds of them. Little yellow eyes. Like tiny cats around a campfire."

I searched the black water.

"Turn on your light," Erin commanded, standing up. I reached into a pocket for a flashlight. There is a limit to how much darkness a person can tolerate at the edge of black water. As soon as I flicked the switch, there they were, staring back at me, pairs of tiny yellow eyes—wary, watching eyes.

The light searched through an underwater world that suddenly appeared, like a green-lit tunnel through black mountains. Down through the water, I could see sea lettuce lifting. Periwinkles spiraling solid against rock. A village of barnacles in a forest of tube worms, with pink palm leaves waving overhead. A fat-headed sculpin fanning lacy fins so closely matched to the sand that they looked like currents stirring the water. Then my light found a pair of little shining eyeballs. The eyeballs bobbled at the base of long, waving antennae, as ephemeral as dotted lines, and the antennae sprouted from the head of a shrimp. There they were, standing on their tiptoes on gravel, little bent cylinders of pink-striped skin around clear water, with beaded legs no thicker than whiskers: hundreds of shrimp, all turned toward us, staring into our lights. In the clean tree-and-rain world up in camp, Frank and our son Jon were leaning over their work in the lantern light, tying salmon flies, unaware of all this gooey flesh, all the striving and scrambling, all these little prickling awarenesses. Erin and I, too: we'd been living our daylight lives on the surface of the island, thinking we were alone out here, if we thought about it at all, thinking we were in charge, never dreaming of the other worlds that come out at night.

"God, I wish René Descartes were here," I said.

Erin laughed; she's used to this. But it was true. Enlightenment philosophers have mapped out such a miserable, lonely world for us to live in. For three, four hundred years, we children of the Enlightenment have sat alone and damp-eyed in a world of nothing but stone and dumb brutes, the only spirit in a universe of matter and mechanical animal-clocks, the only shining eyes in a universe stripped of mystery, exposed to human understanding and control, reduced to human convenience. Lonely kings of the stony mountain, we warily watch the world through our weak and narrow beams of light, denying the existence of everything we can't see clearly and distinctly, the unimagined other worlds watching us in the dark. I wish Descartes would plunk right down here on a slimy rock.

"It's slipperier than you think," I would warn him. He probably would struggle to move around the kelp bed in his long woolen robes—Frank would have to lend him boots—and I know he wouldn't understand my high-school French. But I would clear off a slithery cascade of kelp and point insistently to a rock. Accustomed to moving by candlelight, he might surprise me with how comfortably he finds his way in the dark.

"Here," I would say, grabbing his hand. "Stick your finger into this anemone and feel it shrink away from you. Reach into the water. Try to touch those waving tentacles at the mouth of the tube. See them snap away? Reach out to touch the sculpin; look how it disappears in a puff of sand. Turn off your headlamp. All the eyes will vanish. Move once, just wave an arm, and every sound will stop. And you'll be sitting in a world that is *froid, sombre, muet, et vide.*" Maybe he'd look up startled. "And you, doubting the truth of anything you can't clearly and distinctly perceive, will believe that the world is in fact cold, dark, silent, and empty.

"But if you sit still in the dark, breathing quietly, the world will come to life around you. Astonishment will rise in you like the slow tide, sliding in under the soles of your feet. And then you will understand: you are kin in a family of living things, aware in a world of awareness, alive in a world of lives, breathing as the shrimp breathe, as the kelp breathes, as the water breathes, as the alders breathe, the slow in and out. Except for argon and some nitrogen, every gas that enters your lungs was created by some living creature—oxygen by plankton, carbon dioxide by the hemlocks. Every breath you take weaves you into the fabric of life."

He probably wouldn't understand a word I said.

By then, dampness had seeped through my rain pants, through my fleece pants, into my polypropylene long underwear. This was not good. Clothes dampened by rain or dew will dry again. But once clothes are soaked with seawater, the salt will pull water toward itself incessantly, and the clothes will be irredeemably soggy.

"Ready to go back?" I asked Erin, and we crawled on hands and boots up the slippery rock, like bears, pulling ourselves gingerly over barnacles, until the sharp smell of the sea gave way to the must of ancient cedar duff, telling us we were close to a sip of brandy and a warm sleeping bag.

I'M ONE OF THOSE PEOPLE WHO NEVER CAN THINK OF THE RIGHT words in the heat of the moment. All night I mulled over how complicated and layered and open-ended this kinship of humans with all natural creation actually is, this beautiful, bewildering family.

First, there is the kinship of common substance. Like a sea slug or a horse-neck clam, I am carbon atoms spun through time, arranged and rearranged in patterns. Break my pattern down to atoms, and I can scarcely be distinguished from the stars. Second, there is the kinship of common origins. The gooseflesh that prickles my skin is what's left of the contraction that bristles the fur of a frightened bear and fluffs a bird in February. Third, there is the kinship of interdependence. Consider the sweetrotting hemlocks that create vanillin, which nourishes the underground network of mycorrhizae which produce mushrooms, which feed the flying squirrels, which nourish the spotted owls, which entrance an aging woman, who warms the rotting tree with her fleeced bottom. And fourth, the kinship of a common fate. We, all of us—blue-green algae, galaxies, bear grass, philosophers, and clams— will someday dissipate into vibrating motes. In the end, all of natural creation is only sound and silence moving through space and time, like music.

The same arguments are offered, over and over again, in a dogged effort to preserve a place on the pedestal for humans alone. Here's the most popular one: It would seem that humans are apart from and above the rest of natural creation, for the Bible says that God created man in his own image, several days after He created the birds of the air and the fish of the sea and gave man [sic] dominion over all the creatures that chirp and bubble and roar. This might be a reasonable argument, based on Biblical authority, although scholars quarrel over the meaning of the words translated as "dominion."

Mostly I worry that it's only arrogance that makes humans think that because God made us last He must have made us best. Why couldn't people as readily believe that by the time God got around to making people, He had run out of ideas, so He did a little recycling. There's nothing wrong with learning from experience; that would truly be "intelligent design." The same systems that propel giant squid through the seas move blood through our hearts, and our cells use the same notation that directs the growth of red rock crabs. We breathe the same oxygen as the fish of the sea and the birds of the air, and it turns to the same carbon dioxide in our lungs. To make our minds, our exalted minds, God blew into the brain of a lizard. Our great temples have the proportions of a snail.

It seems to me that if we truly are made in the image of God, then God, too, must be made of the stuff of lizards, or lizards are made of the stuff of God. We should rejoice in that god, and we should rejoice in every human being, and every squid and sculpin—the progenitors of what is divine in us.

A different argument comes from Descartes. Humans have minds, or consciousness, he wrote, "the thinking substance." But plants and animals do not. So humans are apart from and superior to plants and animals.

The fact is, I don't know for sure what animals are thinking, but neither did Descartes, and that seems like a good reason not to rush to judgment about what's on an animal's mind. New evidence emerges every year of sophisticated self-awareness, problem-solving, and communication in animals. Evolutionary biologist Marc Bekoff cites evidence of "dogs that understand unfairness, spiders that display different temperaments, bees that can be trained to be pessimistic," and now animals that have a sense of humor. Koko, the western lowland gorilla who learned sign language, once tied her trainer's shoelaces together and signed "Chase!"

How suspiciously convenient it is to believe that humans have the monopoly of the universe on mind. If people are going to imprison dolphins and transmogrify the gallbladders of bears into fortifying elixirs, if they are going to scrape the bottom of the ocean bare and grind the hindquarters of black-tailed deer into patties, if they are going to reduce owl nesting sites to toilet paper and convince themselves that this is not a problem, then they will need to believe that humans have minds but other animals do not. But this is a matter of convenience, not truth.

Awareness remains the great mystery, unexplained, maybe unexplainable. Which one of us can explain what happens in our own bodies to transform quanta of energy into an awareness of starlight reflected on water? Who knows what happens in a human's mind as she watches that star lift, flare, and tangle in floating kelp—that gladness, that wanting more? And then who could claim to know what happens behind the seeing eye of a shrimp?

Every time we stab our narrowly focused beams of light into the darkness, we find another astonishing world, never imagined. All around us, animals cry out, and wince, and leap, and stare with their bottomless eyes. What pitifully limited imagination would convince us, time after time, that what we see is all there is, that because we have not yet plumbed them, there are no depths?

Then there's a last argument. I don't know if there's a person left on the planet who would defend this argument in public, but I also don't know if there's anyone who doesn't believe it deep down, or hope it's valid. Here it is: Humans can alter the other parts of creation and remain unaltered themselves. Therefore, they must not be part of an interdependent whole.

This is the saddest, most self-destructive mistake of all our sad and self-destructive mistakes, to think that humans can degrade their habitats and not degrade themselves. Counterevidence engulfs us in epidemics of asthma in smoky cities, in lead-poisoned children, in Tlingit stories about the lost salmon, in pandemic disease that strikes hardest on the poor, in broken families and dysfunctional cities, in landslides and vacant streams.

"You could cut off my hand, and I would still live," Powhatan-Ren'pe writer Jack Forbes told me. "You could take out my eyes, and I would still live. Cut off my ears, my nose, cut off my legs, and I could still live. But take away the air, and I die. Take away the sun, and I die. Take away the plants and the animals, and I die. So why would I think my body is more a part of me than the sun and the earth?"

When I sit in the dark and listen, this is what I hear: Seawater drips from sea palms. Rain ticks through hemlocks. A man and his son murmur in lantern light. A winter wren startles from sleep. A kelp crab shuffles under sugar kelp. A boat creaks in the pull of the tide. It's all one symphony—all the small songs sounding in the darkness, coming together to create one beautiful night. Whatever harm we do to any part of the music, we do to the harmony of the whole. The discordance we create will be in our lives and the lives of our children.

Animal behaviorist Marc Bekoff claims that scientists have now agreed that non-human animals have consciousness, citing the Cambridge Declaration on Consciousness (2012): "Convergent evidence indicates that non-human animals have the neuroanatomical, neurochemical, and neurophysiological substrates of conscious states along with the capacity to exhibit intentional behaviors. Consequently, the weight of evidence indicates that humans are not unique in possessing the neurological substrates that generate consciousness. Non-human animals, including all mammals and birds, and many other creatures, including octopuses, also possess these neurological substrates."

(LiveScience.com)

Silence Like Scouring Sand

RAIN POUNDED AGAINST THE RAISED TAILGATE OF MY CAR, where I had taken shelter from the worst of the storm. Water poured from the hemlocks onto the devil's club. From maple bole to bole, raindrops bounced, splattering salmonberry and sorrel. I shrugged into my rain gear, shouldered my pack, and splashed across the parking lot. Rain belled off cars, smacked against my hood, beat on my shoulders, and drummed on the garbage bag covering my pack. Against all instinct, I was going to backpack into the clattering teeth of this North Pacific gale, in search of silence.

It's not easy to find silence in the modern world. If a quiet place is one where you can listen for fifteen minutes in daylight hours without hearing a human-created sound, there are no quiet places left in Europe. There are none east of the Mississippi River. And in the American West? Maybe twelve. One of these is in the temperate rainforest along the Hoh River in Olympic National Park.

At the Hoh River trailhead, where a path disappeared into the gloom under giant, lichen-draped Douglas-fir and western red cedar, I met up with Gordon Hempton. He stood calm and dry under an umbrella, comfortable in the wool and cotton clothes he chooses for their silence. Middle-aged, limber, and weather-tanned, Gordon is a man on a mission to record the natural sounds of the world before they are drowned out by human noise. For years, he has searched for the quiet places where falling water and wren song can still be clearly heard. This weekend, he was taking me to one of the few remaining quiet spots in the United States.

Gordon led me into the dense forest where rain and wind were muffled by moss. Even so, on the path to this silent place, the natural sounds were deafening. "In a forest like this," he said, leaning close to my ear, "a drop of rain may hit twenty times before it reaches the ground, and each impact—against a cedar bough, a vine-maple leaf, a snag—makes its own sound." He crouched beside a fern-banked stream. "You can hear the treble tones," he said, "but do you hear the bass undertones as well?" I knelt on the moss beside him, soaking through the knees of my rain pants.

I had never listened to water quite this way before, with such close attention to its music. "You can change the pitch of a stream by removing a stone." I lifted a cobble out of the water. The chord lost some of its brightness, picked up a drone I hadn't heard before. "A stream tunes itself over time," Gordon said, "tumbling the rocks into place." A channel gouging through the mud that remains after a hillside has been logged is "only noise. But an old mossy stream? That's a fugue." When I had a chance to make sure of the meaning of "fugue," I realized what a great metaphor that was; a fugue is a "musical composition in which a short melody is introduced by one part and successively taken up by others and developed by interweaving the parts." Just like a stream. But Gordon was a great one for metaphors. Once, he told me, he heard wind move up the Hoh valley, knocking dry leaves off the big-leaf maple trees. "It sounded," he says, "like a wave of applause."

For love of sounds like these, Gordon had begun a campaign to protect the silence of the national parks. He called his project "One Square Inch of Silence." Following leads, crisscrossing the country, he searched for one square inch where he could listen for fifteen minutes and not hear a human sound but the whisper of his pencil on wet paper. In Olympic National Park, where 95 percent of the land is protected as wilderness, he found the "widest diversity of soundscapes and the longest periods of natural quiet of any unit within the national park system."

On Earth Day 2005, Gordon marked the site with a small red stone and vowed to defend this tiny space of silence. "Think about finding one place that you can visit, where there will be no trucks heard, no planes flying over, no man-made machinery, no human noise," he said. "Wouldn't that be a beautiful thing?"

It was a powerful idea. As Gordon knew, sound travels. If he could protect the silence of even an inch, he calculated, he would be in effect protecting the natural soundscape of approximately one thousand square miles of surrounding land. It was a first step toward his goal of preventing the extinction of silence.

Past the first milepost, we crossed a glade that borders the Hoh River. After so many days of rain, the Hoh was in full flood. Gray water roared over torn-out root wads, ramming logs against the shore, undercutting the banks, and rolling rocks downstream. Gordon pulled out his sound-level meter, a machine that looks like a handheld radio but measures sound.

"Sixty-three decibels."

The river has the same sound level as ocean surf in a storm. This, Gordon said, is one-tenth the volume of the traffic noise outside the Fifth Avenue entrance to the Seattle Public Library.

Except during the lockdown to slow the COVID-19 virus, cities drown us in sound. Buses grind gears, motorcycles grumble, woofers thud, endless engines combust, trucks beep, and street-corner preachers call down damnation on it all—what does this do to the human being, whose ears evolved as a warning system? In daylight, our eyes can warn us of danger in front of us. But our ears alert us to opportunity and danger twenty-four hours a day, from every direction, even through dense vegetation and total darkness.

When predators are on the prowl, birds and frogs, even insects, fall silent. No wonder humans are drawn to places where the birds feel safe enough to sing. No wonder we smile to hear a frog chorus in the dark. But in the cacophonous city, Gordon believes, we are always on edge, always flinching, the way a deer trembles when it drinks from a noisy river. People continuously assaulted by high levels of traffic noise have suppressed immune systems and significantly increased risk of high blood pressure and heart attacks. A city will be comfortable for a human, Gordon said, only when it's quiet enough that lovers can talk without shouting at each other.

I knew what he meant. I was still on edge from the noise of the drive up I-5 through Portland. When trucks roared past me, they threw rain with such a thud against the windshield that it drowned out even my shirring tires, the smacking wipers, and Willie Nelson's guitar on my radio.

However, it was not noise in the cities that most concerned Gordon, but the extinction of silence in wild places. It was the way that human sounds drown out the music of the natural world that broke his heart—and got his back up. Even national parks are not always able to protect the music of a morning wind in pines, the echo of a woodpecker tapping a hollow trunk, the thrum of distant surf. Dawn in a national park often begins with the brown noise of automobile traffic and swells with the awakening sounds of RV generators, jet overflights, sightseeing helicopters, and "It's a Small World (After All)" playing in the next camp over.

Human noise also damages animals, whose behaviors are exquisitely tuned to songs and other auditory signals they use to hunt and to escape, to establish territories and to find mates. Scientists have documented the harmful effects of mining blasts on elk, passenger jets on bald eagles, the Navy's submarine-hunting sonar on dolphins and whales, and roaring dune buggies on kangaroo rats trying to avoid sidewinder snakes. Just as animals have ecological niches, they have aural niches, defined by the soundscapes they live in. The onslaught of noise destroys that aural habitat. Birdsongs, especially the low-pitched sounds, are lost along highways, which are, in effect, wide swaths of loud, low-pitched noise reaching deep into the forests and meadows, reducing bird habitat and sometimes eliminating it entirely.

It's a loss to humans too. Just as artificial lights drown out the stars, our engines drown out natural sounds, and our experience of the world's beauty is that much more impoverished.

AT 1.4 MILES, WE TURNED OFF THE TRAIL TO CHECK A POSSIBLE campsite, but the hollow where we would have pitched tents was a muddy puddle. We hiked on, splashing along the trail, past dark skunk-cabbage sloughs, over trickling rivulets, under the boughs of cedars so tall their crowns were lost in fog. I was glad for my rubber boots, because the trail had become a stream, and the rocky steps were small waterfalls. We stopped often to listen, putting our ears close to the dark decaying hollow of a stump, or a green carpet of bunchberry, or a fallen log whose crosscut section showed more than three hundred growth rings. Gordon stopped next to an enormous tree and listened intently.

"Do you hear the thrum of the river resonating in the trunks of the Sitka spruce?" he asked. "This is a tree whose wood is chosen for the finest violins."

I tried, but all I heard was the rain and the noise of my own mind, asking mostly, *will I ever be warm or dry again?* and this gray noise, the static that came from my own ears. Gordon was sympathetic. He knew from experience that when people are long enough away from the "chemical whining" of caffeine, aspirin, and alcohol, and from the damage done to their hearing by the noise of the car that brought them out, their ears will silence themselves. And so would my mind.

"Silence is like scouring sand," he said. "When you are quiet, the silence blows against your mind and etches away everything that is soft and unimportant." What is left is what is real: pure awareness and the very hardest questions.

Many years ago, Gordon was a botany student in Wisconsin. As he was driving back to school from the West Coast, night came on, and he stopped to sleep in an Iowa cornfield. Lying on the ground, he listened to crickets scratch their crisp fiddles and corn stalks rasp their leaves. He heard thunder rumble. The crickets went silent, and the storm rolled over him. He heard raindrops smack into soil and hail rattle the stalks. Then the thunder was growling far away, and the crickets were singing again.

How could it be that he had never before heard, really heard, the sounds of the Earth? From that time forward, how could he do anything but listen? How should he live his life? "Whatever came next," he told me, "had to measure up to the honesty of that night."

Now he travels the world, recording sounds with a microphone and creating programs from these recordings—the pure sounds of the natural world, unadorned by human music, uninterrupted by the human voice. His most famous recording project is still probably the "Dawn Chorus," the sound of morning as it scrolls across the planet. It's a remarkable collage of recordings from around the world. As dawn strides across the curve of the slowly turning planet and erases the darkness with light, birds cry out, tentatively at first, insects chime, melting snow strikes stone, a light wind rises, and the whole Earth begins to sing. It is a song of astonishment and gratitude.

Astonishment and gratitude are an important part of what the future stands to lose under the shouting engines of human ambition. When humans silence nature, drowning out the small voices, we subordinate it to our own presumed power. Anyone who has felt the oppression in a classroom or boardroom or marriage when only some are free to speak will understand what it means to be silenced—to have no voice, to be seen and not heard, to be told to "pay attention," which means do not pay attention to any voice but one. Human noise is yet one more oil-fired expression of modernity's claim of sovereignty and control over the natural world.

But silence? Silence creates an opening, an absence of self, which allows the larger world to enter our awareness. It brings us into contact with what is beyond us, its beauty and mystery. Silence is not the absence of sounds, but a way of living in the world—an intentional awareness, an expression of gratitude, to make of one's own ears, one's own body, a sounding board that resonates in its hollow places with the vibrations of the world.

When wind plays across the maple leaves and sets them in motion, it's we who are most deeply moved. No one knows why natural sounds speak so directly to the human spirit, but it's possible to imagine what they say—that we are not separate from the world, not dominant or different. Like stone, like water, like wrens, we carry the shape of the world in our rustling. We are all music, we are all matter in motion, all of us, together sending our harmonies into a black and trembling sky.

AT MILEPOST 2.3, WE SLUNG OFF OUR PACKS, SET UP OUR TENTS, and pulled out a small flask of Scotch. The rain seemed to have let up under the shelter of a Douglas-fir whose trunk was nine feet across. We were not far from One Square Inch. As I tightened the ropes on my tarp, Gordon talked about his efforts to defend the silence of that place. The air is a commons, a public good, shared and cared for by all, he believes. Like secondhand smoke or elevated mercury levels, dumping noise into the air damages the well-being and health of the many, in order to benefit the few—a violation of the ethic of the commons.

THE NEXT MORNING, AT 3.2 MILES, GORDON STEPPED ONTO A path that led through an arch formed by the straddling legs of a great cedar. From there, we followed an elk trail into the woods. We were approaching the One Square Inch of Silence, and Gordon asked only one thing of me. Silence. We walked in—not far, maybe seventy-five yards—through a shallow swale, over shaggy hummocks, to a shoulder-high log fallen so long ago that its bark was coated with moss, and hemlock seedlings had rooted in the duff on its back. There, at N 47°51.959′, W 123°52.221′, was a square red stone that marked the inch of silence.

How shall I describe the beauty of this place? It was an open glade, like the nave of a cathedral, carpeted in deep green moss and deer ferns. There were huckleberry bushes, their bare green branches standing in the rosy litter of their own fallen leaves. The bunchberry leaves had turned red, but the wood sorrel was intensely green. From the forest floor, the columns of the trees rose impossibly high, closing at last in a vaulted green ceiling. Everything glittered with scattering rain. Even the air twinkled, as if it were champagne.

And what did I hear? A tiny lisp—a bushtit maybe. *Tick, tap, pock* of waterdrops, different sounds for every surface they struck. I heard a drop of water pop when it hit a maple leaf forty feet way. There was the faraway rustle of the river. Time passed, unmeasured. Then the quiet filled with the clatter of a bald eagle, a sound like stones shaken in a tin pot. Sitting on his heels in the damp moss, Gordon grinned but didn't speak. A small wind shook a huckleberry bush. A crow called from the crown of an alder. A hemlock needle fell on my shoulder, and I turned, astonished to have heard it land.

Twenty-three hundred times a year, the Navy sends Boeing EA-18G "Growler" jets on electronic warfare training missions over Olympic National Park, including its five wilderness areas. Growlers are some of the loudest jets that the Navy flies. Now the Navy has asked to increase the annual flights to 2,600 per year.

(*High Country News*)

The Song of the Canyon Wren

The love of beauty is a longing for the homeland
of the soul.

—PLOTINUS

THE SONG OF THE CANYON WREN IS THE SOUND OF FALLING WA-
ter. Its bright tones drop off the canyon rim and fall from ledge to ledge a
step at a time, sliding down a pour-off, bouncing onto a sandstone shelf,
then dropping to the next layer of stone-time and down again—a falling
scale, eight tones, a liquid octave of birdsong in the hard, sun-cut canyon. I
lift my binoculars to search the rocks, but I don't find the wren, which won't
surprise you, since you know wrens.

Sometimes sounds turn me almost inside out with longing. The song of
the canyon wren, the faraway voices of my children, and the watery sound
of cottonwoods: These are on my mind today. But the sound of rain on
sandstone will do it too, water hissing at the side of the lost sea, and the soft
breathing of silver fishes caught between grains in this shelf of stone.

I hear singing, and I don't know what to do. I want everyone in the world to hear it. Then I want no one in the world to hear it but me. Then I want to gather my husband and the children and listen together. Then what hits me is a flood of sadness, washing the stones out from under my feet and making me stumble.

Does this happen to other people? It isn't just sounds. It can be a smell, or a glimpse of something in the distance. The silhouette of piñon pines on lavender sunrise sky, or a mountain range under rain clouds, each row of mountains softer and dimmer than the row before, or two black ravens stroking in unison across the red face of a cliff—any of these can hit me a body blow that leaves me gasping. This unnerves me and makes me feel ungrateful. I am blessed by beauty beyond anything I deserve; the gift should make me quiet and glad and at peace, but instead it makes me feel hollow inside. I tell you this, I trust you with this secret, because I suspect sometimes you feel it too.

At first I thought it was loneliness, and maybe that is it. My daughter would love the ravens, I say to myself. Or, if only my son could be here. The beauty is too much for me alone; it opens an empty space that I need to share with someone else, and the absence of the people I love fills me with regret. And maybe it's a vastly deeper loneliness, knowing that even if my daughter were here, or my son, they would never see it the same way. Even Frank, the person closest to me in all the world, sees the land through eyes far different from mine. So I will always be alone in my seeing, fundamentally alone.

Or maybe it's the sadness of unsatisfied greed. I want this for myself, and I want it forever. I am greedy for falling water. Grasping after ravens. Gluttonous, when it comes to cottonwoods. The thought that the sky will dim, that the ravens will land and dive their heads into carrion, that the mountains will disappear into deep night, that the moment—never to be replaced—will be lost to me forever, is more than my greedy little emotional center can support. So maybe this is it: knowing that the moment cannot be captured and held, I mourn the moment as it passes.

Or maybe what I want is to *be* falling water, to merge with the sunrise, to lose myself in the layers of mountains. But I doubt it. So far as I can tell, water falling through sunshine doesn't feel itself falling, doesn't rejoice at the brightness of the light, doesn't know joy or sorrow. If that's true, then to be one with nature would be a pleasure unfelt, which wouldn't be much of a pleasure at all.

A few hours after sunset, the sky glows above the cliff where the moon will rise. Already spires across the canyon are shining white. Bats careen over the water, listening for the echoes of insects. Then the face of the moon rises over the canyon wall and every plane of rock and tumbled stone stands bare and white, outlined by its own moon shadow. I can see Frank in his sleeping bag on the sandstone ledge. A single coyote calls a question and then quiets, listening for an answer. I hear a wood rat scuffing in the dry leaves at the base of the cliff, probably looking for last year's seedpods. A locust in the cottonwood tree suddenly stops buzzing, as if it too is listening. Moonlight has washed most stars out of the sky, leaving only Orion beyond my feet, and the Big Dipper over my head. I lie in my sleeping bag, quiet and alert. What am I listening for?

"Almost everyone is listening for something," Sigurd Olson wrote. "We may not know exactly what it is we are listening for, but we hunt as instinctively for opportunities and places to listen as sick animals look for healing herbs."

Sometimes I think I'm homesick. Sometimes I think that what happens when the landscape seizes me with such sadness is that the moment reminds me of a home I left generations ago, a beloved place I remember in the deepest recesses of my mind. It might be a landscape on an intellectual plane, a Platonic realm of ideas where perfect truth and perfect beauty become one glorious idea that can't be distinguished from love. Or it might be a clean and windswept place, a real place at the edge of water. Maybe something ancient in my mind seeks meaning in the lay of the land, the way a newborn rejoices in the landscape of a familiar face. Maybe I go to the wilderness, again and again, frantically, desperately, because wild places bring me closer to home.

Possibly you think this is all just words, a story I tell myself on dark nights to keep away a greater darkness. It might be so, but there is this fact: Last week, we made camp in a pocket of sand high among sandstone out-croppings above the desert. In the darkness, as stars flickered and the lights of distant campfires flared and fell away, the kettle on the backpacking stove hissed and whirred. But when I turned off the stove, the whirring didn't end. Frank checked the stove for leaks. Finding nothing, he slowly swept his flashlight toward the source of the sound.

In the spotlight was a pale rattlesnake piled like a rope, its tail buzzing, its head probing the darkness toward us. In the whir of the stove and the whir of the snake, I almost understood something. I came so close, but the recognition fell away, after the first startled silence, in the excitement of the snake. In the morning the snake was gone, and there was frost on our sleeping bags. The water in the water bottles had frozen, and the eggs, when Frank cracked them against the frying pan, were as clear and round as glass eyes.

And there is another place. I'll tell you about it, but you shouldn't ex-pect too much, because it's just an ordinary place. We had walked up a dry creek to a clearing where someone had built a windmill. There was a concrete slab and an open basin of thick, green water. The windmill was an ordinary western windmill, held up on steel struts. I sat next to the tank and leaned against a strut. The wind came up warm at my back and the windmill started to creak and knock. Birds came to drink from the basin, including one Frank thought might be a green-tailed towhee. It ran toward the tank like a chipmunk with its tail in the air, but really, there were lots of birds, walking or flitting or diving toward the water. The air was light and warm and, except for the windmill, silent, and the moment was beautiful and true. That is all I can tell you. But I felt I had been given a sudden glimpse of a place I have never been and can only dimly remember.

Because the canyon wren is one of the least studied birds in North America, its population trends are unknown.

(American Bird Conservancy)

How Can I Keep
from Singing?

I. Ghost Lake

GHOST LAKE TRAIL CARVES A PATH THROUGH THE BLAST ZONE where the forest took the full force of the Mount St. Helens eruption forty years ago. Spruce and fir lie where they fell, as if an army of trees, fleeing, had pitched on their faces in the direction of their flight. The blast rolled over them, scorched their limbs, crackled their skin, pressed them into the ash. Rank on rank of shattered tree trunks, half-buried in cinders, reach out with broken limbs. The blast must have flanked the mountain and charged down the hillside, burning the slope to stones. Then the hill slid out from under the trees, dropping them in a heap of pumice and slash. This is where we hiked, my friends and I, across this gray and ruined land toward a pond pierced by snags.

My friends were singing. They dredged their memories for the words to an old hymn. *Through all the tumult and the strife / I hear the music ringing.* When I stopped to listen, I was surprised to hear music too, the pulse of frog song and a Pacific wren chattering in a huckleberry bush. I couldn't see the wren for the new growth of alder saplings and young firs, poking up from the blanket of blueberry shrubs.

An entire forest was rising from the ashes. Silver firs grew higher than my head. Between fallen spars, fireweed and lupine were blowsy with seeds. A goldeneye duck stood on one leg at the edge of the pond. And between the drowned snags, my friend Libby waded barefoot, waving her hand in time with the hymn. *It finds an echo in my soul / How can I keep from singing?*

The sorrow or the song: which is the meaning of the mountain? I'm sure that in one second, only that, the blast seared the eyes of a robin who must have turned in alarm. In the next second, everything was gone but her bones and the silence after the scream. Every bird killed. Every large mammal. Fifty-seven human beings. There are the facts.

At the same time, in the same place, there is the fact of fireweed. There is the fact of frogs blinking by a snowbank behind a hill. There is the coyote, kicking up dust as it defecates a twist of ash. There are the facts of spider silk, green moss, and beavers trooping down new river courses to chew willows nourished by ash and dust.

What is this mountain, where hope rides the back of horror, the two of them galloping in such perfect synchrony that they might be one thing?

II. Horse Ridge

THE TWELVE OF US HAD PITCHED OUR TENTS IN PUMICE ON A north-south spine. Snow-crowned Mt. Adams floated on forests to the east; to the west, the caldera of Mount St. Helens stood raw and gray. Once supper was over and the dishes sort of washed, we sloshed whiskey into tin cups and gathered on the ridge to watch daylight drain from our camp into the bowl of the volcano. Just when we had settled into camp chairs to watch the light flare from orange to red, a rumble of rock sent us scrambling. Rubble bounded down the caldera's western flank. Billows of red dust boiled over the rim, shot through with shafts of light.

I've seen paintings of Creation that look like this—with a red gleam spreading across smoking stone. And I've seen imaginings of the end of time that look just the same—gray rock seamed with fire, rising smoke, falling darkness, nothing left of life except a handful of human beings lined up at the edge of Destruction, whooping.

In my tent, I dreamed about boiling rock and roused myself to make sure I was still as cold as I thought I was. Not long after, a loud rumbling in the dark shook me out of sleep. Light strobed the side of the tent. An eruption? I was on my knees at the door. But it was lightning that crashed between the mountains. Thunder rolled again, and I unzipped the tent. The broken side of Mount St. Helens filled the doorway. A bolt of lightning shot from the crater almost to the moon, half hidden behind storm clouds. Light flooded the tent. Thunder shook the ground under my knees. Should I run down the road to the landing and bring up a van to protect us, all my friends who chose their tent sites for the view? Making camp on the highest ridge for miles around had seemed like such a good idea. I counted off seconds to gauge the distance of the storm and turned my head to judge its direction. The lightning might pass to the north, I decided. It might not.

In the morning, I pulled on a rain slicker and climbed out into a cold, wet day. Dawn flooded under dark clouds that still dashed lightning onto the mountains to the north. Rain rivered down the road. But sunshine poured over clouds on Mt. Adams, and when we gathered for breakfast, a meadowlark sang.

Under the flapping tarp that sheltered the picnic table, a geologist hung onto the tent poles to keep the whole structure from winging across the valley. With his free hand, he gestured wildly toward the crater. *The gas cloud over the 1980 eruption was full of lightning,* he said, *zapping up, down, every which way. It charged the nitrogen in the ammonia and ozone. So the ash that fell from the cloud was full of nitrogen that fertilized new plants. Cool, huh?*

He could hardly contain his excitement at the ecological possibilities of the eruption. And in fact, all the scientists there were jazzed by what they call this great "biomass disturbance event." Disturbance kicks the world into motion. What had been stagnant is booted into new life. Mature stands of fir and pine are replaced by meadows and thickets and creek bottoms. What had been simple is now complex. What had been monochrome forest is now a hillside of purple lupine. What had been silent is noisy with chatter and song.

I was not ready for this. It unnerved me to be confused about the difference between destruction and creation. *Destruction, creation, catastrophe, renewal, sorrow, and joy are merely human ways of seeing, human projections onto the landscape*, the ecologist insisted. *What is real*, he said, *is change. What is necessary*, he said, *is change.* Suddenly, there are new niches, new places to grow and flourish. In ponds, landslides, rocky hillsides, and a great profusion of edges, beetles troop over stumps, huckleberries emerge from snow banks, astonished pocket gophers dig into the sun. Between the forest patches, meadows fill with the creatures of wet prairies and oak savannahs.

This "biomass disturbance" has made a new place for birdsong. I understood that. But at what terrible cost to nestlings burned in their nests, the sudden silence where there had been hungry peeping? Lord knows, I wanted to see the world the way ecologists do. If all of us thought of death as change rather than catastrophe, we could blunt the edge of sorrow. And isn't this a source of hope, that the forces of nature turn death into life again and again, unceasing? *Above Earth's lamentation / I hear the sweet though far-off hymn / That hails a new creation.*

That's how the song goes. But sometimes I can't hear the hymn for the crying of small birds in the back of my mind.

III. The Trail from Donnybrook

THOSE SCIENTISTS WERE SO WRONG BACK IN 1980. WHEN THEY climbed from the helicopters, holding bandanas over their faces to filter ash from the Mount St. Helens eruption, they did not think they would live long enough to see life restored to the blast zone. Every tree was stripped gray, every ridgeline buried in cinders, every stream clogged with toppled trees and ash. If anything would grow here again, they thought, its spore and seed would have to drift in from the edges of the devastation, long dry miles across a plain of cinders and ash. The scientists could imagine spiders on silk parachutes drifting over the rubble plain, a single samara spinning into the shade of a pumice stone. It was harder to imagine the time required for flourishing to return to the mountain, all the dusty centuries.

But there they were: on the mountain only thirty years later, these same scientists on their knees, running their hands over beds of moss below lupine in lavish bloom. Tracks of mice and foxes wandered along a stream, and here, beside a ten-foot silver fir, a coyote's scat grew mushrooms. What the scientists knew then, but didn't understand before, is that when the mountain blasted ash and rock across the landscape, the devastation passed over some small places hidden in the lee of rocks and trees. Here, a bed of moss and deer-fern under a rotting log. There, under a boulder, a patch of pearly everlasting and the tunnel to a vole's musty nest. Between stones in a buried stream, a slick of algae and clustered dragonfly larvae. "Refugia," they call them: places of safety where life endures. From the refugia, mice and toads emerged onto the blasted plain. Grasses spread, strawberries sent out runners. From a thousand, ten thousand, maybe countless small places of enduring life, forests and meadows returned to the mountain.

I had seen this happen.

So I was careful when I talked to my students. They had been taught, as they deserve to have been taught, that the fossil-fueled industrial growth economy has brought the world to the edge of catastrophe. They were fully informed about the decimation of plant and animal species, the poisons, the growing deserts and spreading famine, rising oceans and melting ice. If it's true that we can't destroy our habitats without destroying our lives, as Rachel Carson said, and if it's true that we are in the process of laying waste the planet, then our ways of living will come to an end—some way or another, sooner or later, gradually or catastrophically—and some new way of life will begin. What are we supposed to do? students asked me. What is there to hope for at the end of this time? Why bother trying to patch up the world while so many others seem intent on wrecking it?

These are terrifying questions for an old professor, and it took me some years to think of what to say.

I decided to tell them about the volcano. I told them about the rotted stump that sheltered spider eggs, about a cupped cliff that saved a fern, about all the other refugia that brought life back so quickly to the mountain. If destructive forces are building under our lives, then our work in this time and place, I told them, is to create refugia of the imagination. Refugia, places where ideas are sheltered and encouraged to grow.

Even now, I said, we can create small pockets of flourishing, and we can make ourselves into overhanging rock ledges to protect their life, so that the full measure of possibility can spread and reseed the world. Doesn't matter what it is, I told my students; if it's generous to life, imagine it into existence. Create a bicycle cooperative, a seed-sharing community, a wildlife sanctuary on the hill below the church. Tear out the irrigation system and plant native grass. Imagine water pumps. Imagine a community garden in the Kmart parking lot. Learn to cook with the full power of the sun at noon.

We don't have to start from scratch. We can restore pockets of flourishing life-ways that have been damaged over time. Breach a dam. Plant a riverbank. Vote for schools. Introduce the neighbors to each other's children. Celebrate the solstice. Slow a river course with a fallen log. Clean the grocery carts out of the stream.

Maybe most effective of all, I told them, we can protect refugia that already exist: they are all around us. Protect the marshy ditch behind the mall. Work to ban poisons from the edges of the road. Save the hedges in your neighborhood. Boycott what you don't believe in. Refuse to participate in what is wrong. There is hope in this, an attention that notices and celebrates thriving where it occurs, a conscience that refuses to destroy it.

This is integrity, I told my students, which is wholeness, which is healthy, which is holy. This is consistency of belief and action. And that is the answer to hopelessness: to do what you think is right, knowing that your actions will be the wellspring of the new world. You'll know you have achieved this integrity and torn loose from hypocrisy, I told them, because the relief of it will bring you to tears.

From these sheltered pockets of moral imagining, and from the protected pockets of flourishing, new ways of living will spread across the land, across the salt plains and beetle-killed forests. Here is how we will start anew: Not from the edges over centuries of invasion. Rather, from small pockets of good work, shaped by an understanding that all life is interdependent, driven by the one gift humans have that belongs to no other— practical imagination, the ability to imagine that things can be different from what they are now. *My life flows on in endless song*, the hymn says. *How can I keep from singing?*

IV. Birth of the Lake Trail

I DON'T KNOW WHAT I WOULD SAY TO THE STUDENTS IF I WERE to meet with them now. I have left the university. It moved ponderously; it paid attention to unimportant things; it was thoroughly stuck in self-perpetuating privilege; it punished ideas that offended donors; it invested its endowment in death and buildings; it wasted my time. Worse, it squandered its extraordinary, maybe unique, potential as a world-saving refugium for wisdom, experimentation, and courageous questioning.

Among the subjects I once taught was inductive logic, including the wondrous argument by analogy. If two entities are similar in many ways we can observe, it may be that they are similar in ways we can't observe. The blasted volcanic plain and the Earth itself are similar in many ways: both are moved by tectonic forces over long periods of time; both have been devastated by agents of death, the volcanic explosion and the Sixth Extinction; both are seeded by resilience. So if the blasted volcanic plain is blooming again, sooner than we ever expected, then perhaps the Earth also will be reseeded by good works, recover quickly from the Sixth Extinction, and bloom again.

But an argument by analogy is quickly weakened, often lethally, by dissimilarities. As far as we know, the Mount St. Helens explosion did not drive any species to extinction. The mountain returned immediately to its preexplosion climate. No powerful industries are swarming over the mountain yet, mining and drilling and poisoning it to dust. In fact, the mountain is now a federally protected park. None of these are true of the battered Earth. This might reasonably shake a person's confidence in a quick recovery.

I have no doubt that life on Earth will come roaring back once the destruction has ended, just as it did on Mount St. Helens. Climate catastrophe and the Sixth Extinction will create one hell of a "biomass disturbance." Just as on Mount St. Helens, the surviving species will rush to fill the disturbed ecosystems and suddenly empty niches. Some species will thrive in the dramatically different climates; others will not make it through. I don't know if humans will be among the survivors. I know that many songbirds, insects, amphibians, and human children will not. I expect that the Earth will, for millennia, be a far quieter place. In that quiet, for better or for worse, those beings who listen may *catch the sweet though far-off hymn / that hails a new creation.*

To date, there are 202,467 wildlife refuges and protected areas on the planet. These protect 20 million square kilometers, which amount to 14.7 percent of Earth's land and 10 percent of its waters. Entomologist E. O. Wilson argues that in order to save Earth's biodiversity of lives, fully half of the planet must be protected.

(International Union for the Conservation of Nature,
Half Earth)

4.

SING OUT

LISTEN: THE SCHOOLKIDS ARE SHOUTING AS THEY MARCH UP Thirteenth Street. A bullhorn calls out, "We are the future!" The children answer, "Give us a chance!" As passing cars honk, the children flinch, then cheer. Up Washington Street, students from the university march behind the blats of a brass band. When the schoolkids and the college kids meet at the intersection, everybody hurrahs and the groups merge—part of 6 million people worldwide. "You have stolen my dreams," some of the little ones shout. But then who can hear the words over the singing crowd and tuba? Marchers pump their signs up and down. Grownups on bikes wobble as they match the pace of the parade and, from the windows of a parking garage, a hundred office workers cheer. Police sirens whoop. At the courthouse steps, the crowd quiets to listen as a microphone squeals and a teenager begins to shout, "We will keep fighting until the fossil fuel companies are held responsible for their crimes." In small, fluting voices, the children call out, "How dare you? How dare you?"

After the Fire, Silence and a Raven

I'D ALWAYS THOUGHT THAT WHEN I DIED, MY CHILDREN WOULD scatter my ashes under the pines beside Davis Lake, where we have camped for twenty years. But it's all ashes there now. Wildfire flared up along the road behind the east campground. Driven by twenty-five-mile-an-hour winds, the firestorm charged across the road and ran hard and fast through miles of lodgepole pines, torching the manzanita and exploding into the crowns of the trees. Fire crews pulled back and let it burn. What else could they do, with the fire roaring around them?

For two weeks, heavy smoke and unpredictable winds closed the roads into Davis Lake. Helpless, I watched on the internet as the fire expanded, red spots encircling the lake until the whole map was blotchy red. But as soon as fire crews cleared the trees that had fallen across the road and removed the barricades, I headed for Davis Lake, driving the Cascades Highway through a healthy forest of ponderosa pines.

Car windows down, I can smell the thick beds of pine needles, warm and sweet in the sun. Ponderosas gold as honey rise from meadows of green grass, blue sagebrush, currants already turning orange. Then suddenly the color is gone. For miles in front of me, all I see is a blanket of white ash, stuck through with tree trunks, broken and black, and the shadow of a raven swerving between spars. A thin line of smoke rises from a smoldering stump. I pull into an overlook on the side of the road, step into ashes, and listen.

I had loved the sound of Davis Lake in the spring. I remember waking early one morning, years ago, in my sleeping bag under pines at the edge of the lake. As Frank and our little ones breathed quietly beside me, great blue herons flapped over the marshland, croaking. Red-breasted nuthatches called from the pines. Coots splashed in circles, and sandhill cranes clattered on the far side of the lake, leaping and flapping their wings in a clumsy dance. I remember how the frogs shouted that morning, filling the air like a cheering crowd. In a pine far down the lake, fledgling eaglets begged without ceasing, a scraping sound like pebbles against steel. I had settled deeper in my sleeping bag, warm and grateful.

But now the music is gone, and the silence is so complete that I brush my ears to be sure I can still hear. Finally a raven calls. A single grasshopper scratches in black stubble. The wind lifts slender whirlwinds from the ashes. But without pine needles to make music of the wind, even the whirlwinds are silent. I stand with my head back and my eyes closed, trying to understand how it could be so suddenly gone, the green singing life of this place.

I'm a philosopher by trade, so I should know how to be philosophical about loss. The world is in flux, and change is the only constant. Forests are no exception. They grow and burn and grow again. I know this. Everybody knows it. Almost three thousand years ago, the Greek philosopher Heraclitus acknowledged the necessity of change: you can't step into the same river twice, he said. But why not, I want to know. Why can't what is beautiful last forever?

Everything has to change, Heraclitus answered, because all the world is fire and water in constant conflict. Fire advances and is quenched by water. Water floods and is boiled away by fire. And so people wake and sleep, live and die, the fires of their spirits steaming against the dampness of their flesh.

Summer changes into winter, as sun gives way to rain. The mountains boil up from the seas, and the seas come into being and pass away. Forests are reduced to ashes, and from the ashes rise new forests, damp and shining.

How easy it is to write these words, so good in theory. But in fact, the only thing rising from the ashes today are the whirlwinds. A pickup rumbles into the overlook. A man steps out, inhales sharply, then turns to his buddy. "Look at those huge big dirt-devils," he says. The two of them stand without speaking, watching ashes lift in spiraling threads and flatten against the sun.

ONE AUGUST EVENING MANY YEARS AGO, FRANK AND I CROUCHED on the beach with our children, thrilled and terrified. We flinched each time lightning struck the forested ridge three miles across Davis Lake. A thunderous crack, a flash of light that turned our eyelids blue. Then a flame flickered on the ridge and a tendril of smoke rose into the air. Lightning struck into the forest again and again until the hillside was dotted with little flames, each with its trail of smoke, like candles on a birthday cake.

Over our shoulders, the moon rose, flaky and red. The lightning moved slowly away over the lava ridge, flashing silently above the eastern plains. Frank tucked the children into their sleeping bags, then sat by the tent, watching across the lake. I launched a canoe onto water that pooled red around my bow. Flames shot streaks from the far side. Every pull of my paddle spun off a ruddy spiral and vibrated the lake into red and purple ripples. I could smell smoke and water, damp algae on the shore. As I rocked in my boat, black clouds drifted over the face of the moon, and water licked in the reeds. Soft rain fell, ticking on water that faded from red to gray, and one by one, those fires went out.

Water won that round. But I knew that fire's time would come.

For eighty years, lodgepole pines have grown up thick as doghair on the flats around Davis Lake. A person deliberately laying a fire in that forest couldn't have done a better job than the trees did themselves: Pile kindling under each tree, stacks of downed branches, hard and silver and scratchy. Sprinkle the kindling with dry pine needles. Drape branches into the kindling so any ground fire is sure to climb the tree. Let the forest bake in the sun and the drying wind. Then all it takes is a spark—dry lightning, a Bic lighter, an ATV.

Lodgepole pines need to burn. It's the earliest science lesson I remember.

When I was growing up in Cleveland, somebody mailed my father a shoebox crammed with lodgepole pinecones. Although he had never seen a lodgepole forest, he had read of these western trees in biology books and the journals of Lewis and Clark. So as my sisters and I jostled around him, he spread the cones on a cookie sheet and baked them in the oven. We watched through the oven door, marveling as the cones bloomed like roses in the heat, releasing papery seeds.

My father handed each of us a seed, and together we admired this wonder: how the cones stay on the trees for years, tight as fists, until fire warms the resin that holds them shut and they release the seeds to replant the burned-out forest, a forest made in just such a way that the very fire that destroys it will create it again.

So I can understand the battles between water and fire, cycles of living and dying, the urgent necessity of death, all of us designed to die—just exactly that—everything we love designed to die, as the lodgepole pine forest is made to burn. I can understand this in my mind, but how will understanding ease this loss?

You can say that it's all a natural process, that the appearances and disappearances are all the result of cycles working themselves out over time. A lodgepole pine forest isn't merely the place that shelters my tent. It's a process of growth and change, trees transforming themselves to ashes to green seedlings to barren spars to seeds on the wind. The lake isn't only a place where my children float with blue dragonflies. It's a stream of water that flows from light on the mountaintops to the long, dark caves, emerging into the blue lake and plunging into the dark again like a serpent that has no end.

You can say it's something particularly human, this tendency to misunderstand natural change as unsupportable loss. You can say that sorrow is part of the same arrogance, the same self-centeredness that leads humans to measure time by the span of their own lives, to define what is real by their own needs.

If I could step outside my own life span and purposes, then maybe I could make myself believe that the difference between a natural disaster and a natural cycle is only a matter of time. Aldo Leopold advised his readers to think like a mountain, on that timescale. A mountain wouldn't mourn the loss of a forest any more or less than a human mourns the leaves that twist off an oak in autumn.

But how can I think like a mountain? Tell me: Does a mountain feel its scree slipping to its feet?

ALL THE BACK ROADS INTO DAVIS LAKE ARE CLOSED BY BARRI-cades and striped tape. But the assistant fire manager for the Deschutes National Forest, Gary Morehead, agrees to drive me in. I pull on the fire-resistant yellow suit he hands me. Then he shoves his truck into gear and steers it around the first barricade.

We follow a dusty track through a landscape of black tree trunks stuck at every angle to the ashes. Here's where bulldozers scraped a clearing in the forest, Gary tells me, a safety zone where firefighters could retreat if the flames turned on them. Here's where the force of the fire-generated wind snapped every tree and sent it crashing into flames. The truck bumps down to the site of the campground, next to the creek that feeds the lake. As I step out of the truck, my feet lift smudges of dust.

Here in the ashes is a fire ring, solid and ironic, and the bent frame of a lawn chair, tossed on its head. A sign that once pointed to the boat landing is reduced to a bolt stuck through a post burned to a spindle.

The fire created a wind so fierce, Gary says, that it threw a canoe into a tree and drove flames against it until the canoe melted over the branches. When the branches burned away, all that remained was a metal frame collapsed against what was left of the trunk. The fire vaporized two spotted owl nests. Too young to fly, the young owls surely died. And the firestorm sucked the eagles' nest out of the tree. No one can find the fledglings.

I trudge along the edge of the lake. Where, in ashes turned violet by the ferocity of the fire, is the place where I lay on pine needles with our newborn baby, pointing out yellow-rumped warblers and chickadees? I want to find the shallow reed bed where Jonathan, grown into a toddler, waded after minnows; before we could convince him to leave the water, his legs were streaked with leeches. I wanted to find the place where Erin wove tule reeds into a little fort, crawled in, and read Dr. Seuss, tracking the words with her finger. Could it be here, in this empty space, that Frank and I drank wine at a picnic table, talking about our kids gone off to college until stars popped out, spangling in the pines like lights on a Christmas tree?

One year this ground was covered with toads. Another year, it was baby garter snakes and buttercups. Now, the ground is covered with black stubble burned right to the water. I stand beside the cove where Frank and I rode the canoe back to shore on a windy day, streams of light and water hitting our faces. I remember the cool wind, the bucking canoe, the exhilaration, the wheeling eagles, but all I can see is the lake calmly reflecting the devastation of this place, just sitting there, as if all that life, all those precious, irreplaceable times hadn't roared into flames and vanished forever.

I ASK GARY IF WE CAN GO TO THE HEADWATERS OF RANGER Creek, where I remember a spring that flowed out of the mountainside and made its way through flowered meadows to Davis Lake. He's reluctant to go there, not sure if the place has been secured. But he circles the truck around the yellow tape and brings us slowly through the ashes to what was once a willow flat at a bend in the creek.

I'm surprised by what I see. Stumps are still smoldering, two weeks after the fire. But already, new grass has grown four inches high, green as frogs. The willow thickets have burned down to blackened stubs that reach up like a hand extending from the ground, the black fingers as short as my own. But inside each hand, as if spiraling from a wound in the palm, is new growth, the coiled leaves unfurling. There are birds here, osprey, soaring over the water, watching for trout.

The springs surge from the barren ground into a burst of green, shimmering and miraculous in that field of gray. Trees have fallen over the creek and burned in from both ends, but between the banks, the tree trunks are intact, shading the water, shaded themselves by tall green rushes. Wherever water reached into the ashes, plants grow. Rosy spirea, yellow cinquefoil, sweet bracken fern.

What is this world, that it has all these things?—the dead and dying forest, the charred bones of young eagles, and water pouring from the earth, ancient snow finally emerging again and flowing into the great expanse of blue. What is this world, that life and death can merge so perfectly that even though I search at the edge of water, I can't find the place where death ends and life begins?

I am standing here, in all my color, the blue veins in my elbows, the reddened skin on my knuckles, my fireman's yellow suit; standing here in all my noise, the breath in and out, the wind flapping my collar. But some day my children will bring my ashes, gray and silent, to be caught up in a dirt-devil that makes no noise at all. And where will the color be then, and the sound of a person breathing? This silence, so hard to understand.

In Heraclitus's world of constant change, don't we all yearn for some pause in the river, an eddy, where the water slows and circles back upstream for a long, calm time before it rejoins the flow? This is what Davis Lake was for me, a quiet circle of the seasons, a place where the world seemed to come to rest. A place my family could return to, year after year, as the cranes returned, as the water returned, and the yellow blooms of the bitterbrush. The constancy of the lake had reassured me, the reliable circle of life.

But in this greening place of ashes and springs, I begin to understand that time cannot move in a circle, coming again to where it was before. Time sweeps in a spiral, going round and round again—the cycles of the seasons, the flow of the cold springs, the growth of a forest or a child—but never returns to the same place.

And shouldn't I be grateful for this, that birds will nest in the Davis Lake basin, even though that particular pair of owlets will never fly again? Trees will grow beside the creek, while my grandchildren play on the green-banked stream. Willow thickets will tremble with morning ice, the songs of red-winged blackbirds, the slow unfolding of next year's dragonfly's wings. And we who love this world will tremble with the beauty of the spiral that has brought us here and the mystery of the spiral that will carry us away.

The 2019–2020 wildfires in Australia were the worst on record. Midway through the fire season, 10 million hectares had burned, an area the size of South Korea. Thirty people were killed, a billion wild animals killed, tens of thousands of cattle killed, and air quality reached twenty times "hazardous." Scientists blame record-setting heat and persistent drought, both attributed to climate change.

(Carbon Brief, Yale Environment 360)

We Will Emerge Full-Throated from the Dark Shelter of Our Despair

THE DAWN CHORUS

WHO AMONG US HAS NOT SEEN THE IMAGE OF THE EARTH FROM space, as the planet turns under the light of the sun, and morning advances? Always half spangled black, half glistening blue and green, the ball rolls, and mountains and coastlines, seas and continents slowly emerge into the sun. It stirs me, I have to say, to watch satellite video of morning flowing over the Earth like a bright tide. I imagine a flotilla of boats full of brass choirs, bumping along the advancing edge of the flow, heralding the morning.

But in fact, that's almost exactly what has happened every day for, how many days? Sixty million years, so . . . doing the math . . . 21,900,000,000 days. For that many days, there have been birds that stir in the breeze that precedes the day, lift their heads, and begin to sing. The "dawn chorus," people call it, the sequence of especially brilliant bird songs that begin the day.

Who knows why birds pour out their hearts first thing in the morning? Maybe the air is cool and sound travels farther. Maybe other noises haven't yet filled the sky. Maybe the morning, before the foraging gets good, is spare time that the birds fill with singing, the avian analogue of a second cup of coffee. Probably singing in the morning demonstrates fitness, saying,

"I am strong, I am fully alive, I have lived through the night and emerged full-throated from my dark shelter with energy and joy to spare." All good reasons to sing.

Apparently, birds wait for the light that will tell them it's safe to sing, and then the songs burst out of them. Studies in Ecuador show that birds who roost in the tops of the trees generally sing first, maybe because that's where the light hits first. It takes the shadowed birds a bit longer. Species wait their turns, each entering at its chosen place in the fugue.

I can identify where I have lived by my memories of what bird sang first in the morning. When I was growing up in Ohio, the mourning doves sang first, pouring song into the dense peony-scented summer air. In graduate school in Colorado, I was up before the birds, always doing my assigned reading at the last possible moment, but I was glad for the company of robins, when they finally woke up in the ponderosa forest. In Alaska, the sapsuckers are first to drum in the morning; but if you don't count drumming as song, then it's the whistle of the varied thrush that wakes us. In Oregon, robins were the first to sing, but the Eurasian collared dove, an invasive species, now beats them out. Yesterday, away from home, it was a rooster crowing and the roaring departure of flight 820 to Seattle that woke me long before light.

Recording artists have followed the sun around the world, recording the dawn chorus on continent after continent, the great unrolling of song as the Earth turns: again, that tide of sparkling light advancing over the darkness, lifting song after song.

Imagine the magic of this planet. Dear god, we live on a music box. No, we live on a player piano, the rotating cylinder plinking out music as the Earth turns. No, we live on something even more magical than a music box or a player piano, because pink morning light, not little metal tabs, plucks the strings. Who could have the whimsy and the aesthetic abandon to invent such a marvelous machine? What parent would not treasure it and pass it down, carefully preserved, to her heirs?

As it happens—how reluctantly I take this turn, but there's no denying—we are losing the music box. No, that's wrong: human expansion is destroying the music box. It would be difficult to record the world's dawn chorus now, only twenty years after the first recordings, for two reasons. One is the increasing din of anthropogenic noise. For creatures that sing to live, noise destroys habitat as surely and as efficiently as bulldozers. The other reason is the steep reduction in the number of singers in the choir and the total elimination of some of the voices. I hate this.

If the dawn chorus weren't so splendid, and surely one of the unique wonders of the universe, maybe its destruction would be bearable. Or if its destruction weren't so completely avoidable with a little care and humility and even appreciation, maybe its loss would be bearable. I don't know what is worse: to do terrible harm purposely or with complete oblivion. They both seem unforgivable.

The only thing that keeps me from despair is knowing that there is another kind of dawn chorus circling the planet. It is not shrinking; it is growing. It is not quietening; its voice is rising to a roar. The story of the other dawn chorus came to me from my friend, a woman in Alaska.

MY FRIEND COULD NOT SLEEP. EXACTLY: HOW CAN ANY OF US sleep? There is so much work to be done to save habitats, to save democracy, to save decency, to save children, and yes, to save birdsong. We could work all the bright day and all through the night, and still the work would not be done, and how can it even begin? My friend lay awake in a darkened room, one hand gripping the other. Who can let herself fall asleep, when she has not found a way to save the world? Because that's the task ahead of us, all of us. Is this not so?

Every evening, she watched as bare branches and telephone lines sliced the falling sun as if it were an egg yolk, and the day darkened, until one night she remembered that the sun was, of course, not falling. The far edge of the Earth was rising. As each of us falls into bed at night, exhausted and despondent because our work is barely begun, the sun is rising on the other side of the planet. Other people are rising to the challenge of protecting what is flourishing and just and beautiful.

On the rotating planet, there's a great dawn chorus of committed people, millions and millions of them, who rise from their beds or mats or blankets, rustle up coffee or atole or tea, and set off to do the good work of defending the world's thriving. We can hear the chorus if we listen. The rustle, the creak of doors made of tin or wood or grass, voices calling out to each other in a thousand languages, the roar of action advancing around the world, awakened like birds by the rising sun.

When night comes now, my friend is able to sleep, and when the dawn comes, she takes up her part of the work that others, exhausted, have laid down. Because my friend is Alaskan singer-songwriter Libby Roderick, her part of the work is to write the songs. Here are some of the words she wrote in "The Cradle of Dawn":

> Sunset in your country, sunrise in mine
> Lay down your body, hear mine begin to rise.
> Sunset in my country, sunrise in yours
> I feel you there in the dawn. . . .
> There are no promises that we will see the day
> The dreams we live for will succeed.
> But I can promise you that halfway round the world
> I'll hold the light up while you sleep.

Each of us will emerge full-throated from the dark shelter of our private despair. We will find our cause. We will find our courage. We will find our chorus. Our work now is in the streets, in the state houses, on the riverbanks, in the college quad, in the churches. What we cannot do alone, we can do together.

IN THE REEFS OFF WESTERN AUSTRALIA, THERE IS A DAWN CHO- rus of fishes. Every day, as the sea brightens from black to azure blue, the fish raise their voices. To call in a mate, to gather the pack for the morning hunt, to defend their territories, to greet the sun, who knows? Maybe to rejoice that they lived through the night. The blackspotted croaker moans like a foghorn. Tiger perch grunt. The batfish goes *ba ba ba*. When all these solo- ists join their voices in green slanting light, they sing a great dawn chorus.

Let starfish dance *en pointe*. Let spinner dolphins pirouette. Let stick- lebacks do the zigzag wag, while birds sing out, the brazen sun rises like courage, and all who love the world go off to the work of saving it.

September 2019 saw the largest climate protests in world his- tory. An estimated 4 million people from 163 countries partic- ipated in 2,500 events on all the world's continents, including Antarctica.

(Vox.com)

The Sound of Mountains Melting

A BLOCK OF GLACIER ICE HURLS ITSELF FROM THE DEPTHS OF the seawater into the clear air, massive as a breaching whale. Kittiwakes flap and scatter, screaming. When the ice crashes down, its impact throws a wing of water fifty feet up the face of the glacier. The water falls back in a thunderous torrent and the ice block surges up again, but not so high as before. Then again, the ice crashes down, the water sluices up, the kittiwakes scream. The violence of rising and falling, the thunder and echo, the scatter and glare—in all this, the block of ice bounces and then settles, finding how it will float as an ice floe on the sea.

Kittiwakes circle madly, shrieking, *kittiwake kittiwake*, diving to feed on small shrimp stunned by the violence. Our skiff rises on a slow swell.

I never knew that a tidewater glacier could calve underwater. But it makes sense that the ice front, above the waterline and below it, would lurch forward and disintegrate at the same time. And when a subsurface block of ice is unlocked from the glacier, of course it will shoot up into light and bird-cries. I never thought about how the impact, ice and water come to blows, would affect the birds and their prey. I expected the thunder. I am surprised by the cries.

But no time to wonder. A crack sharp as rifle fire, then the grumble of tumbling ice. We flinch and swivel to watch chunks bounce down an ice slope. Suddenly undermined, the pinnacle above it begins to slide. The slab gains speed, then the whole thing disappears under a descending cloud of snow. The avalanche hits the sea and suddenly all the force of down, down is heading back up.

The boom, when it comes, is a bombshell. We are drifting a hundred yards from falling bombs, the shock and awe, birds like shrapnel. The face of the glacier is not smooth, like an office tower; it is all shattered pinnacles and spires, broken and leaning, veined with rubble, more like a cathedral town after a war.

The icefalls are unpredictable, and our nerves are on edge. The sun has been heavy on the glacier, and nighttime temperatures and even the sea temperatures have been unusually warm. Each day, across the entire face of the mile-wide glacier, a mass of snow and ice forty feet thick falls off and melts in the sea. I am sitting in a skiff on a rolling swell in dazzling sun, smelling salt, squinting at sea-level rise.

How beautiful sea-level rise is today. It feels strange to say, but it's true. Saltwater: smooth and thick as green-bottle glass, reflecting icebergs and kittiwakes. Sky: intensely blue and without a cloud. Sunlight: a shattered disco ball, shooting white spears. The air is baked solid with silence, until it blows up, when the booms and gull cries echo in countless ice caves and crevasses. The Hoonah people, whose ancestors lived here hundreds of years ago, before the glacier surged over their land, come back to dance on land exposed by melting glaciers. In a circle, the women sit still, watchful. A man strikes a skin drum. The women rise up, waving their arms and shouting, repeating a word that sounds like *kittiwake. Kittiwake, kittiwake.* Here is the ecstasy of danger. *Kittiwake, kittiwake, kittiwake.*

The beauty of the falling ice signals disaster. The more glaciers and ice sheets melt, the higher sea levels rise; the higher sea levels rise, the more saltwater inundates farmers' fields, the more storm tides drive people from their homes, the more villages are taken by the sea, entire island nations submerged: around the world, a rising tide of dislocation and suffering.

We should not be surprised by the paradox of beautiful danger; the world dazzles us with danger, time after time. When dry winds scour dust from desert playas, sunsets are never more ablaze, clouds of dirt on scarlet fire. When lightning strikes a hillside in the night, the trees spark like meteor showers. In the light that streams under the purple clouds of an advancing hurricane, every tree and field glows green, and a red barn, soon to be rubble, is never redder. The beauty of a melting glacier is even more menacing. It signals not a moment of change, but a million years; not a place destroyed, but a world.

Hank cranks the key that starts the outboard and turns our bow into the swell. This is Hank Lentfer, our lanky friend, showing us his beloved Glacier Bay. He grew up next to the tidewater glaciers and explored from one to the next, skiing across the snowfields at their heads, climbing the cirques to map the shear-line of the glaciers in the trees. Now he is mapping the sounds of the Bay, bird calls and whispers, waterfalls, the huff of nervous bears, the click of mountain goat hooves on rock. If anyone should be brokenhearted at the melting, the prospect of Glacier Bay without glaciers, that would be Hank. I ask him how he holds beauty in his heart, as it melts away. He shushes me with a raised hand. He is video-recording.

I sit and listen, but the quiet fills my mind with our friend William. William was an irrepressibly happy man. He had a smile that made you think he was about to spring a terrific surprise, happiness held in, waiting for the moment. Last month, climate change killed him. He was our friend, and climate change killed him. Under the stalled jet stream, hard rains and wind had pounded Southeast Alaska. Four inches one day. Six more the next. On a mountainside near Sitka, wind uprooted ancient spruce and rain loosened mud and rocks. Tons of muddy debris roared down, engulfing a house under construction. William was inspecting the wiring on the new house. Two minutes later, he was lying curled and dead under that pile of mud and rocks.

Three hundred thousand climate-change deaths every year and rising. Every life as sacred as the next. Every grief the same scream and the same long horror. I glance at Hank and find that he is watching me. His hand gesture tells me, just listen. Sit quietly and listen.

Light travels faster than sound. So first the gull-rise. Then the ice chunks trickle down, all the little facets, falling. The icefall undermines the spires. Two great towers fall on their faces, still gaining speed when they hit the sea, and then, only then, the screams reach us, the clatter and rumble, the crack, and the boom. The calving has created an echo chamber for its own thunder, which rolls and rolls until the roll of the sea reaches the boat and we rise like inhaled breath and fall again.

It is impossible to be quiet in the face of this glacier's spectacular collapse. *Aaah*, I say and then remember that Hank has signaled me to be still, a silent awe. But how is that possible? *Awe* is onomatopoetic; its sound is its meaning. What better word for that mix of fear, excitement, and appreciation than a body's response, that exhalation, that *awe*?

It is getting late, although the sun will not set for hours. But as it rolls along the peaks of the mountains, the sun is losing its color and the air chills, and we began to think about finding a place to camp. It will be a perfect campsite, with a view down a fjord to the glacier where the sun will finally hide its face.

Avoiding the biggest ice chunks, Hank slaloms the skiff through deepening sky and a skim of slush like clouds. Waves have eroded the icebergs into fantastical shapes: a swan with an arching neck, a fist raised from a bloated body, a shelf of sliding books, ducks—lots of ducks—and, unaccountably, a ballerina. Holes pock many of the icebergs; each cradles a dark stone. Most of the blocks have dark stripes. "Landslides spilling off mountains onto the ice," Hank says. The bottoms of the floes are wave-bashed into sharp facets, smoothed into curves, or cut into a lattice of caves. One berg looks like a barge carrying a load of gravel.

Hank steers beside a baby seal who stares up from a chunk of ice no bigger than a kitchen sink. Even as we stare back at her, she holds her ground, for some reason unwilling to slide into the safety of the sea. And then a killer whale's black fin shows against an iceberg not so far away. He patrols slowly past the baby, then turns to glide back the way he has come.

The island where we decide to camp was born around the time I was born, when the glacier stalled and melted, dropping polished stones in a long ridge. There hasn't been time for trees to grow, but the boulders are laced with runners from beach strawberries and blanketed by dryas mats. In a pocket of sand, dwarf willows have begun to grow. Here, beside them, we smooth away the pawprints of a brown bear and pitch our tent.

A little slosh of whiskey, a small beach fire. The sea darkens and the tide begins to flow, carrying a parade of icebergs past camp. Kittiwakes fly to a roost on a nearby rock face, a quieter ruckus now, *ock ock ock*. Occasionally, a crack rings out as another chunk of ice breaks from the glacier's distant face. We know that in a half hour or so, a tiny tsunami will push a riffle a few inches up the beach. A raven hops toward the fire and leaps away, muttering.

I am muttering too.

Of all the things that the brutal, extractive fossil-fuel economy takes from us, it is now poised to take away the simple enjoyment of beauty. I feel guilty when I admire the glory of the decaying face of the glacier. I feel like an accomplice when I *ooo* and *aaah* as it cracks apart. It feels wrong to be so happy here, not just witnessing but cheering on the great unraveling that will flood homes and fields. It feels wrong to admit that this is probably the most beautiful set of events I have ever seen, even as I know that fathers on the other side of the planet are lifting wailing children onto their shoulders and wading through pestilential mud, away from the only homes the children have ever known, and with no refuge in sight.

Not just guilty. Beauty makes me sad now. When I watch the shorebirds in a sandflat in front of me—those little sanderlings, buzzing along the shore, poking their bills into the mud, birds lifted all in a cloud that flicks from black to white as the flock turns into the light and away again—all I can think of is 80 percent. Eighty percent reduction in the number of sanderlings in the United States since 1970. I start to count the sanderlings. Are there twenty, when there should be a hundred?

Nineteen.

My heart skips a beat. I see a cloud of dazzle above a rising humpback whale, and even as I call out in surprise, I start counting down to 57 percent, the percentage of humpbacks that remain. I'm not sure my grandchildren will be able to watch sanderlings or spouting whales in Alaska when they reach middle age. I am quite sure they may never see a calving glacier.

Our essayist friend, Scott Russell Sanders, writes that there is a connection between what is beautiful and what is good.

> What we find beautiful accords with our most profound sense of how things ought to be. Ordinarily we live in a tension between our perceptions and our desires. When we encounter beauty, that tension vanishes, and outward and inward images agree.

But in a stripped-down, struggling world, what we find beautiful may be in profound discordance with how we think things ought to be. All is not right with the world. Its beauty deceives us. Taken from us now is the idea, as old as Plato, that there is any necessary connection between what is beautiful and what is good. Photographs in a Central American newspaper showed cumulus clouds reaching to the heavens, and beneath them, brown earth sprouting automobiles and bringing forth shattered window frames.

"Born in clouds worthy of Michelangelo," the newspaper said, "the floods in Central America have killed more than 10,000 people."

Taken from us also is the idea that there is a connection between what is beautiful and what is safe. We find open savannahs beautiful, ethologists argue, because our ancestors recognized the safety in the long, uninterrupted view. We find chubby, bright-eyed children beautiful, because their faces signal robust good health. And so it is with beautiful adult faces: the bilateral symmetry, that perfection, is a sign of an easy birth and unmarred growth.

These are confirming examples of the connection between beauty and safety. But counterexamples abound. And now I am thinking again of the beauty of a calving ice sheet, a danger to boats that get too close, and a danger to everything too close to the rising sea. Its beauty doesn't make it dangerous; its danger doesn't make it beautiful. It is just both. The island is glossy and new, the willows are glossy and new, the beach is glossy and new, the whole world smells like a newborn baby, salty and milky like that. But only because the glacier is melting. It is a sign of both the glorious creativity of the ancient Earth and the reckless iniquity of the human beings who are on track to destroy it. It's just both.

Night descends on our camp as rapidly as the moon rises, and soon the edge of water picks up the light and wanders onshore, as if it were a hundred lamplighters offering a flame to each black stone. I feel that my heart would stop if I lost all this, and the terrible longing is a mystery to me. What is beauty saying so insistently?

The world is profligate with beauty, as it has always been and will continue to be, scattering beauty about recklessly, as if there were no limit to its supply. The world makes beauty—it is helpless to do otherwise—and drops it on the back of a fish, places it carefully on a bird's beak, buries it in the sand, embeds it in layers of cambium under cottonwood bark, casts it to the winds, throws it like sparks into the sea, fires up auroras in the night sky when everyone is asleep. It makes no empty promises of safety or salvation. Beauty came before us, and beauty will be here when we are gone.

"You know, Kathy," Hank says, "this is all the world's beauty asks of us—to notice it, to be glad for it."

I carefully put down my little cup of hooch and honey, taking time to balance it on a stone. But I knock the cup off kilter, and the spilled honey also is beautiful, a sheen on the rising tide.

By 2100, 36 percent of the glaciers in the Himalayan mountains and the Hindu Kush will have melted, even if the world warms by "only" 1.5°C. The melting will critically affect the water supplies of 2 billion people, including the 250 million who live in the mountains, and 1.65 billion more who rely on rivers flowing from the Himalayan glaciers into India, Pakistan, and China.

(*The Guardian*)

Rachel's Wood Pewee

ON WONDER

I HAVE BEEN TRYING TO LEARN TO IDENTIFY BIRDS BY THEIR calls. You'd think this would be easy. I can pick up a tune quickly and, like it or not, remember it forever. Why else would Everly Brothers songs endlessly repeat in my mind? But bird songs just don't seem to stick to me. Frank, for all his other virtues, couldn't pick up a tune if it had a rope handle. But he can walk along a trail and call out the names of the birds by their songs. I don't get it.

I can read the music of a flock of blackbirds on a five-strand barbed-wire fence. Any singer can do it. Take their little black bodies to be musical notes; take the wires to be a musical staff, the five horizontal lines and four spaces that represent musical pitch. The position of the birds, arrayed high or low on the lines, mark the melody. If you can read music, you can sing birds on a fence, although sometimes they line up one above another, and you need friends to help sing the chords. But this ability doesn't help me much when what I want to do is remember which song which bird sings.

Just last week I was reading Rachel Carson, the environmental writer who changed history with her book *Silent Spring*. This is the book that

sounded the alarm about the deadly effect of the pesticide DDT on songbirds. I imagine she called the book *Silent* Spring, rather than *Colorless*
Spring, or *Odorless* Spring, or *Tasteless* Spring, or Spring *You Cannot Feel*,
because it was the loss of the birds' *music* that would grieve her the most.
Her love for birdsong saturates her books. Hear how closely and fondly she
listened:

> And then there were the sounds of other, smaller birds—the rattling call
> of the kingfisher that perched, between forays after fish, on the posts of
> the dock; . . . the redstarts that foraged in the birches on the hill behind
> the cabin and forever, it seemed to me, asked each other the way to Wis-
> casset, for I could easily twist their syllables into the query, "Which is
> Wiscasset? Which is Wiscasset?"

My mother tried to teach me bird calls that way, translating their calls
into words. Birds make surprising statements in English. American goldfinches say "Potato chip." Barred owls ask "Who cooks for you?" My mother
said that eastern towhees say "Drink your tea"—good British advice. But I
understand that American ornithologists translate the same towhee call as
"Hot dog. Pickle." You'd think the ornithologists were hungry. But maybe
they had other needs beyond food: The hermit thrush says "Why don'tcha
come to me? Here I am right near you." The solitary vireo calls "Come here
Jimmy quickly." But then we're back to the presumed hunger of the Mac-
Gillivray's warbler, "chip-chewy-chew."

These mnemonics help. Somehow it's easier for me to remember words,
spelled out, than the slippery, wavering, vanishing calls. Then it's just a matter of remembering who says what. You've got to love the birds who say
their own names and get it over with. Killdeer. Chickadee. Cuckoo. Willett.
Poorwill. Kittiwake.

Just yesterday, I learned the most wonderful thing about paying attention to birds. Searching through Rachel Carson's papers in the Beinecke

Rare Book and Manuscript Library at Yale, a writer named Maria Popova discovered a single sheet of stationery from the Portland Rose railroad company. At the top of the page, Rachel had written, in scrawling pencil notes, *pee-a-wee—a wee*. The page below was filled with rows of the strangest hieroglyphs. Short, shaped lines, each maybe half an inch, one after another: *Down-up. Up-all the way down-up. Up-down. Down-up. Forward slash. Up-down-up. Down-up. Up-down with a little bend at the end. Down-up.* One can imagine Rachel sitting on a log, jotting down her transcription of the song of the wood pewee. It made an intriguing pattern. How closely she must have listened. How joyfully she would have made the little marks, tracing the path of bird songs on the page. How deeply she must have wondered about the meaning of the marks she made.

I imagine she felt the same wonder listening to the wood pewees as she felt listening to sandpipers and oystercatchers at the edge of the ocean. "Contemplating the teeming life of the shore, we have an uneasy sense of the communication of some universal truth that lies just beyond our grasp," Carson wrote in *The Edge of the Sea*. "What truth is expressed by the legions of [animals]?" I would like to know too. What are they saying to one another, and what would their mingled voices tell us if we listened?

Maybe it's not so important to learn the names of the birds after all, I'm starting to think. Maybe it's a mistake to put English words in their mouths. Maybe we don't even need to know who's saying what. Maybe we just need to listen.

"It is not half so important to *know* as to *feel*," Rachel wrote. You don't have to know the names of the plants and animals to nurture a sense of wonder. "Drink in the beauty, and think and wonder at the meaning of what you see." And then Rachel told of her plan to take her nephew into the garden to look for the Fairy Bell Ringer.

She had heard a small *ting* in her garden some nights before, a *ting* in the midst of the cricket calls and owls. "It is exactly the sound that should come from a bell held in the hand of the tiniest elf," she wrote, "inexpressibly clear and silvery, so faint, so barely-to-be-heard that you hold your breath as you bend closer to the green glades from which the fairy chirping comes." She and the little boy took flashlights and crept through the damp grass, but they never found who rang the fairy bell. That would remain a mystery to wonder at.

Rachel hoped to write a book about wonder but had time in her abbreviated life only to write an article for the *Woman's Home Companion*, "Help Your Child to Wonder." HarperCollins later published it as a book called *The Sense of Wonder*. I love this book.

The Sense of Wonder begins on the Maine seacoast at night, in the rain, "just at the edge of where-we-couldn't-see." Of the sky, nothing is visible. Of the sea, only dimly seen white shapes. Carson and her little nephew laugh for pure joy, sharing the "spine-tingling response to the vast, roaring ocean and the wild night around us." Carson calls a sense of wonder an emotion, but the philosopher in me wants to use the old-fashioned word *passion* and its old-fashioned meaning—when, moved by some outside force, a person feels and responds. But what outside force, what feelings especially, what response?

Wonder is a sudden surprise of the soul, René Descartes wrote. "When the first encounter with some object surprises us, and we judge it to be new, or very different from what we knew in the past or what we supposed it was going to be, this makes us wonder and be astonished at it." *Astonish*, from the Latin *tonus*, thunder, to be struck as by lightning, the sudden flash that startles us and, just for a moment, lights the world with uncommon clarity, the honest awe that Christopher Sartwell calls "the shock of the real."

The emotion that comes next could be delight, if a person is struck by the beauty of nature or the brilliance of its design. But what catches someone by surprise can take her into a darker place too, where she is clobbered by the sweeping wing of the deeply mysterious beyond the boundaries of human experience. Then wonder leads to "a sense of lonely distances," Rachel admits, a sense of isolation from what is profoundly apart. Loneliness turns to yearning, love for something beautiful and mysterious and other.

Rachel compared wonder to a child's view of the world, where everything is new, and the child is open to a surprise around every corner. The late rabbi Abraham Heschel called this seeing "radical amazement."

> Wonder is a state of mind in which . . . nothing is taken for granted. Each thing is a surprise, *being is unbelievable*. We are amazed at seeing anything at all; amazed . . . at the fact that there is being at all. . . . Amazed beyond words . . .
>
> Souls that are focused and do not falter at first sight, falling back on words and ready-made notions with which the memory is replete, can behold the mountains as if they were gestures of exaltation. To them, all sight is suddenness.

It seems to me that if wonder is the capacity to see as if for the first time, then wonder has a moral purpose. It's a lot like the moral purpose John Dewey found in art: "to do away with the scales that keep the eye from seeing, tear away the veils due to wont and custom, perfect the power to perceive," and "enter . . . into other forms of relationship and participation than our own." That makes me think: Maybe that kind of attention is necessary for any moral relationship—the ability to set aside our own stories and recognize and sympathetically listen to the story told by someone else. If so, then a sense of wonder is the open eyes, the sympathetic imagination, the respectfully listening ears, seeking out the story told by nature's rough wings and flitting wrens and, by that listening, entering into a moral relationship with the natural world.

That close relationship is a source of strength, healing, and renewal, Rachel believed. "Those who contemplate the beauty of the earth find reserves of strength that will endure as long as life lasts," she wrote. "There is something infinitely healing in the repeated refrains of nature—the assurance that dawn comes after night, and spring after the winter."

If I could, I would tell Rachel a story about our grandson and a red-winged blackbird. She would understand why the story means so much to me. For reasons I still don't understand, the little boy was dejected on the day we hiked to a marsh in a bird refuge. He sat crooked and slumped on a rail fence, turning away from a red-winged blackbird that balanced on a cattail nearby. *Okalee.* This child is our whistling grandson, but he didn't respond; maybe he didn't even hear the call. Undaunted, the bird continued to sing, *Okalee. Okalee?* Finally, the boy lifted his head, as if he couldn't help himself, and listened closely. *Okalee,* he whistled to the blackbird. *Okalee,* it whistled back. *Okalee. Okalee.* "I think that bird is answering me," he said, and it surely seemed to be. Their confidences went on for a long, wondrous time.

Wonder is the opposite of boredom, indifference, or exhaustion. "If I had influence with the good fairy who is supposed to preside over the christening of all children," Rachel wrote, "I should ask that her gift to each child in the world be a sense of wonder so indestructible that it would last throughout life, as an unfailing antidote against the boredom and disenchantment of later years, the sterile preoccupation with things that are artificial, the alienation from the sources of our strength."

I think of "sterile preoccupations" and marvel that Carson could have so clearly foreseen our own time, sixty years away. The economic forces of our lives are centripetal, tending to spin us in smaller and smaller circles, creating a kind of solipsism that comes from separation from the natural world and our biocultural communities. It's not that we humans aren't natural creatures; it's not that we don't live always in the most intimate contact with the natural world that seeps in our pores and rushes through our blood. It's that we sometimes lose track of that fact or deny it, and so shut ourselves off from a large part of our own humanity. When we measure our successes and failures against our own mean interests, they grow to grotesque proportions. Then, self-importance and self-absorption bloat and distort our lives and our relationships.

Meanwhile, Earth turns, birds fly north or south, fish rise or sink in the currents, the moon spills light on snow or sand, and we—do we think we turn the crank that spins the Earth? A good dose of wonder, a night of roaring waves, a faceful of stars, the kick in the pants of an infinite universe, the huge unknowing—these remind us that there is beauty we didn't create. There is mystery we cannot fathom. There are interests that are not our own. There is time we cannot measure. When we live humbly in full awareness of the astonishing fact that we have any place at all in such a world, we live richer, deeper lives, more fully realizing our humanity.

And so a sense of wonder impels us to act respectfully in the world. There is meaning and significance in these products of time and rock and water, far beyond their usefulness to human purposes. The sweep of time and the operations of chance have created a world that leaves us delighted and dazed, struggling to understand the very fact of it, its colors, its squeaks and songs. It deserves respect, which is to say that a sense of wonder leads us to celebrate and honor the Earth.

Too, a sense of wonder shows us our own responsibilities to care for the objects of wonder—to do them no harm, to protect their thriving. Rachel: "Wonder and humility are wholesome emotions, and they do not exist side by side with a lust for destruction." To the extent that she's right, wonder may be the keystone virtue in our time of reckless destruction, a source of decency and hope and restraint.

Lately, I've been studying the scream of the red-tailed hawk. So ferocious, so reckless, it has become to me "a symbol that stood for life itself," as Rachel put it, "the delicate, destructive, yet incredibly vital force that somehow holds its place amid the harsh realities of the inorganic world . . ."

I am proud that I can distinguish the hawk's screech from the Steller's jay's, which is a perversely spot-on imitation, so the last time I was out in our field with our grandsons, I quite confidently pointed into the oaks on the fence line and announced to one of the boys, "Hear that? That's the red-tailed hawk." He almost fell over laughing and pointed to his brother, who was blowing a sharp call through a blade of grass held tight between his thumbs. The laughing, screeching boys, the blade of grass, the red-tailed hawk, the Steller's jay, Rachel's redstarts and pewees—all of them call out with the same eagerness for life, all in their own languages, haunting and untranslatable. Maybe by listening to them with a wondering heart, we can, in Rachel's words, "approach the ultimate mystery of Life itself."

When the U.S. government chose the bald eagle as the national bird in 1782, the country held about 100,000 nesting birds. But almost two hundred years later, a combination of hunting, habitat loss, and poisoning by DDT had reduced the numbers to an estimated 487. After the United States outlawed hunting eagles and banned DDT, the eagles recovered to 9,789, about 10 percent of their original abundance. The U.S. Fish & Wildlife Service now recommends tripling the number of bald eagles that the electricity-generating industry can kill without penalty to 4,200 per year.

(U.S. Fish and Wildlife Service, FWS.gov)

Hear the Wind Blow

EVENING AT THE EDGE OF THE SEA. BLUE CLOUDS HEAPED AT the end of the inlet, low sun gleaming on the tide flats. We're sitting on an assortment of chairs at the edge of a porch, maybe twenty feet above the beach. There are a bunch of us, wearing wool hats and holding paper cups of salmonberry wine. The old-timey bluegrass band is tuning up. It takes a while because they are laughing about something, these four long-haired guys in rubber boots and sweatshirts. No hurry, because the audience has lots of tuning up to do too, coming into the harbor with stories to tell and another gallon of berry wine fermenting in the shed.

There aren't enough chairs. There are never enough chairs, so people sit on overturned buckets and benches made of halved logs. But there is plenty of food on the planks of the potluck table. There will be broiled shrimp from the handsome young crabber; baked beans in molasses from the harbormaster; lots of potato salads, because everybody grows potatoes; and rum cake baked on board by one of a half-dozen stout, white-bearded fishermen, indistinguishable except for the variety of their guffaws. The children line up beside the planks, studying the food as if they expect it to hatch.

The band finally converges on A, the banjo rips off a mighty lick, the people rotate on their buckets to face the music, and they are off, double-time. *Blow ye winds in the morning, blow ye winds high-o.* Each parent reaches out an arm and hauls in a kid. *Clear away your running gear and blow, boys, blow,* and this is perfect: flag swaying, feet stomping, yellow sun shafting in, salt-scent rising, clouds sagging dead on the water as if they had been pierced by cannon fire. The convergence doesn't last, of course. One of the kids quietly spins out of his dad's grasp and slips away. Then the exodus of the children begins. Big kids, middle-sized kids, and little ones clomping away in oversized hand-me-down boots, ribbons sailing from their hair. Soon we can hear them calling out as they run on the beach below us, and this gives a joyous contrapuntal kick to the madly picked mandolin. Oh, the harmonies are tight, as the music and the breeze gust in wind-channels across blue-gray bays.

From my stump-seat at the edge of the porch, I have a great view of the guitar and the kids, and I could swear the kids are running to the rhythm of the guitar as they race up and down the beach. They are, of course, running with sticks, but these are harbor kids, and not a parent calls a warning. If they can't keep from stabbing somebody or putting out an eye, they can't make it on a fishing boat. They roll big stones and poke with sticks, and now they are hooting and jostling. Only one kid stands apart in a blue sweatshirt, a lanky little kid, twisting one foot into the other.

The bass player winks at his girlfriend and launches the next song. She beams back. *Goin' up Cripple Creek, goin' at a run. Goin' up Cripple Creek, have a little fun.* But the child apart is yelling. I can just make out his reedy voice: "Stop it! You guys, stop it!" He takes a deep breath and now he is running into the clump of kids, grabbing something and heaving it overhand like a baseball into the sea. Another kid chases him, but he grabs something else and throws it into the drink before a little girl grabs his throwing arm and holds on tight.

It takes a while before he wrestles free and runs up the bank and into his father's arms. It takes even longer for me to figure this all out, the way he pants and points. And now here is an older boy, disputing every detail.

"Justin stabbed a beach crab through with a stick, stuck it to the ground, and then all the other kids started sticking crabs to the sand."

"Nuh-uh. Everybody was trying to find the biggest crab, but Justin's crab pinched him, so he stabbed it with a stick."

"Those crabs were innocent. They didn't do anything wrong."

"The kids only stabbed the males, and there's way too many male crabs anyway."

"I tried to throw the crabs in the ocean to save them, but Chrissy held me down."

"Yeah, because you stole Chrissy's crab and threw it in the ocean."

"Because she was going to *kill* it."

"It wasn't a stick anyway. Justin had an iron rod."

The banjo squeals all the way up the D string and swings into another song. *I asked my love to take a walk, to take a walk, just a little walk. Down beside where the water flows.* The crowd sighs and sings along. Sitting there on the stump, watching the boy cling to his father in the galloping music, breathing baked beans and rock wrack, I try to make some meaning out of all this.

But there are no metaphors here. The older boy tells me later, "The moral of the story is, don't stab crabs when a tattletale is around." In a different story, the crab-thrower might tell anybody who says his action made no difference that it made a difference to the crab that was saved. But there is no moral to this story. There are only facts: Children are strong enough to impale crabs with sticks. Children are strong enough to throw crabs into the sea to save them from the first children. The more children impale crabs, the more children join the slaughter. The more children throw crabs into the safety of the sea, the more other children will do the same. I don't know if there are two kinds of children. I suspect that each child is capable of each thing. The beach exhales in horror and holds its breath in hope. *Hang your head over, hear the wind blow.*

Adults are as strong as children. So every person is strong enough to decide to save the animals from the cruelty. This is a choice every person can make. But this also is true: For all the people who decide to save an animal, there will be others who try to block them. It's important to know this will happen. Other people will try to hold them down. Others will laugh at them. The winds will be against them. Cruel winds ride a long straight fetch through history. It takes moral courage to push against their force.

When a fearsome storm is bearing down on a great ship—the first winds shuddering in the stays, the first waves burying the bow, sullen clouds obscuring the horizon—the captain shouts the order. "All hands on deck." Every seaman knows what that means. Each person on board, no matter their rank or watch, has an absolute duty to rush from their gambling tables or bunks to their stations, to do whatever has to be done to save the ship.

What if seamen don't respond? "Give me a minute, I've got some lucky dice for once." Or "wake me if it gets really bad." Under the old laws of the sea, that would be a flogging offense, or worse.

The analogy is a harsh one. The climate disruptions that are bearing down on the planet—intense heat, failed crops, waves of desperate refugees, extinctions, and acidified seas—are a planetary emergency. Unaddressed, they are likely to take down the ship. "All we have to do to leave a ruined world to our grandchildren," wrote Yale dean Gus Speth, "is to keep doing what we are doing now." This is a call for all hands on deck. But the order is not coming from any captain. In the face of the perfidy of reality-denying government "leaders," the call is coming from all quarters: Indigenous people worldwide, scientists, religious leaders, human-rights activists, national security advisors, economists, parents, and local government officials. Turning back this crisis, to the extent that it is still possible, will take the greatest and most determined public collective action that the planet has ever seen.

The danger is that inattentive citizens might not step up to help, not enough. And that is a moral failure; let us say it straight. Those who stand aside are taking advantage of the actions, often sacrifices, of those who step up to demand or offer solutions. If the children of the inattentive have fresh drinking water, if their grandchildren have enough to eat, if their coastline property fends off the rising seas, it will be because of the courage of others, not their own. But the inattentive are not just doing nothing. Their silence reinforces the message that this climate disruption is no big deal—exactly the message the fossil-fuel industries and their government minions want to convey. In that way, those who fail to respond to the emergency call, distracted or dozing, become part of the storm itself.

Climate change is a call to all hands to rush on deck, to help save what they care about the most by doing what they do best. Is it writing? Public speaking? Singing? Organizing? Is it walking in a parade? Is it even blocking a bulldozer? There might have been a time when our work for the world was quiet work in our private lives, focused on exemplary living and careful consumption. That time has passed. Our work now is in the streets, in the state houses, in the college quad, in the grocery line, speaking out. Speaking out against the corporate plunder of the planet. Raising our voices to defend future generations, plants and animals, the desperate poor, the children, even the crabs.

Hear the wind blow through the open door where the musicians lean against the jam. *Hang your head over, hear the wind blow.* Hear the wind blow in the beach grass, rasping the blades as if it were sharpening them. Hear the wind blow from the snowfield on the mountain, a silver sound, and cold. Hear the wind blow from the low-pressure area sinking over central Alaska, dragging in squalls. Hear the wind blow down the long fetch of the inlet, threatening thirty knots by midnight. Hear the wind whistle up the storms. Hear the wind blow from the beginning of time to the end of time. Everything depends on the courage to do the right thing and, if courage fails, to call for help. People of the sea know this well.

In 2018, crabbers in Oregon reported only 25 percent of their usual catch. The same decline was expected in northern California. The Pacific Coast Federation of Fishermen's Associations sued fossil fuel producers, including Chevron, ExxonMobil, and BP America, alleging that they have knowingly contributed to climate change for decades, damaging their livelihoods.

(Oregon Dungeness Crab Commission, *New York Times*)

Be the Bear

I. Glacier Bay

"BEAR!" THAT'S WHAT YOU SAY IN ALASKA WHEN YOU SEE A BROWN bear trundling toward you down the beach. Not "Run!" That's a bad idea.

She was midsized, maybe three hundred pounds, brown with a golden saddle. Not running, exactly, but not wasting any time either. The distance between us was closing at a speed that struck me as alarming. Trucking along behind her was a yearling cub. The little guy was having trouble keeping up. He stumbled along, whining.

We had been throwing spinners at the mouth of a narrow stream, hoping for trout or even a salmon. It was bear country, Glacier Bay, Alaska, so we weren't surprised to see a bear. But we were surprised that she was heading for us, rather than away. We had been making lots of noise, and there were five of us plus an aluminum skiff pulled up on the bank. That's a formidable gathering, and we would have expected a mother bear to slip silently into the brush and disappear.

"Yep, bear," I said, as I reeled in and moved behind the skiff.

"I don't get this," muttered Frank. "She knows we're here."

The bears continued to make a beeline toward us, clambering over a log next to a willow swale, splashing through a tidepool. Every so often, the sow glanced over her shoulder into the alder forest on the hill that rimmed the beach. Following her glance, we saw what she was running from. A line of wolves, maybe five of them, fifty feet or so up the hillside, stalking her cub. They darted in and out of the trees to keep her in sight, never faster than the bear, but never slower, making no sound at all. One wolf was silver, the others dark and mottled. Ravens were yelling by now, and we could hear the bears' paws crunching gravel.

When she came to the edge of the stream just across from us, the sow stopped and lowered her great head, swinging it back and forth. Now she'll turn away, I thought. She didn't. All five of us were backing away, keeping the boat between us and the bear, but there was a tangle of alders behind us, a limit to how far we could retreat. She waded into the water, with the cub following. It was a narrow stream, maybe thirty feet across, with a deep pool in the center. The sow and the cub stopped there. She was close enough that I could clearly see water drops beaded on her nose. There, she began to do . . . what looked a lot like play. Floating as upright as if she had been wearing an inner tube, the sow reached over and dunked the cub. He sputtered and grabbed for her shoulder. She swatted him under again. They hugged and rolled, chuttered and hummed. We stood frozen on the bank, watching the bears goof around. That's what they were doing—goofing around. The wolves had melted into the forest. We did not see them again.

After a time, the bears shambled up the stream bed to a sandbar, where the sow dug in the sand and the cub fussed around. When we looked again, they were gone. For a long time, I sat on the beach and sifted sand through my fingers, thinking this through. That was a brave bear, to come close enough to us to scrape off the wolves. They wouldn't have harmed the sow, but they surely would have killed her cub.

I am the daughter of a mother who was ferocious in her children's defense. She had the courage to stand up to anything that threatened her children. I remember a feckless school principal, a nest of wasps, a perverted Methodist minister, even once a striped skunk. She fought them off, risking friendships, a hundred stings, nasty backlash, and stink, to prevent a greater harm. When a fight was over, I remember, she would cry.

I am also the mother of two children. I understand why the bear would risk everything, even a crazy-close approach to five people and a motorboat, to keep her cub safe from what she assessed was a lethal threat.

When our children were babies, I cradled them in my arms and vowed to keep them safe. Isn't this what all mothers do? But our children are not safe, as the relentless drilling and burning of fossil fuels breaks down the planet's life-sustaining systems. Several years ago, five hundred scientists, led by a team from Stanford University, warned the world: "Unless all nations take immediate action, by the time today's children are middle-aged, the life-support systems of the planet will be irretrievably damaged."

Let's not deceive ourselves: that damaged planet will not be safe for the little ones. The children need a healthy world that provides them a healthy life, not a poisoned, bulldozed, desperate world, where forests burn and rivers run dry. Not an insane, nihilistic, unstable world where powerful people pursue profit, even to the end of civilization. What else is my work, my role-responsibility, the duty that comes with loving the children, if not to defend their future?

I would like to be that bear, with her cunning and her courage. I would like to be the bear, with powers honed, sharpened, and passed down mother to daughter over generations. I will be my mother, who loved me and protected me and taught me to stand up for what I care about more than anything else in the world.

II. Southeast Alaska

TWO CHERISHED FRIENDS HAVE COME TO VISIT US IN OUR ALASKA cabin. They are Libby, a singer and songwriter from Anchorage, and Robin, a botanist and writer from New York. The three of us decided we would go for a walk. This is how lots of stories start, isn't it? Three small succulent beings setting off down the path, whistling.

At the base of the path from our cabin, we turned right, past a salt marsh of sedges where bears often grazed, and entered the trail that circles the cove. Alders arched over the trail, creating a tunnel of green leaves spangled with light from the glare of high tide. The path was moss and mud, a quiet path, a narrow path that forced us to walk single file. So we talked as we walked, raising our voices to a volume that would be rude in anyone's living room, and in fact seemed rude on the trail, but never mind. I remember we were talking about partner songs, two completely different songs that create wonderful harmonies when they are sung one on top of the other. Like *When the Saints Go Marching In* and *Swing Low, Sweet Chariot*. Or *Skip to My Lou* and *Rock'a My Soul*. But it wasn't as if we weren't paying attention; we watched the trail ahead, called out around the corners, and, in every mud hole, we scanned for bear tracks.

But there were no tracks, and we turned the corner to climb up mossy hummocks through Sitka spruce and hemlocks to a low headland. There, the trail broke open to the inlet, a bright surprise after the dim, enclosing forest. We paused there, listening to waves break on the rocks. A flock of turnstones pecked along the beach below us, conversing in a language made of rattles and skirrs.

"I don't feel comfortable going any farther," Robin said. "I feel that the bears are not welcoming us."

Robin is a member of the Bear Clan of the Potawatomi people. I know no one more attentive to the moods and inclinations of the beings in the forest. So no questions. No hesitations. We turned on our heels and began to walk back the way we had come. Down we went over the hummocks, through the alder tunnel, under splashes of light. In a few more yards, just past the sedge swale, we would reach the base of the path to our cabin. Then we would turn left and climb up to the safety of its porch.

There was a bear feeding in the sedges. A big one. His back was turned to us. Either he had not heard us coming or he didn't care. We could always retreat, but there was no place for us to go but back into the forest where we were not welcome. We could keep going toward the path to the cabin, but that meant approaching within a few yards of the bear.

We huddled to think this through. The bear turned his massive head and glanced over his shoulder at us. We backed away a few steps, grabbing hold of each other for courage. The bear swung around to face us. Standing straight, Robin began to speak to the bear in the Potawatomi language. She must have spoken sentences, but I couldn't separate them into words; they were more like a river than a cobble beach. But there was no need to translate; the tone of her voice said clearly, *We mean you no harm. May we pass safely? We mean you no harm.* The bear lifted his head and turned his eyes away.

Shall I sing to him? Robin whispered. *I know his song.* She did. She sang. A deeply rhythmic song, a chant maybe, calling to the bear in her strong voice. The bear made no visible response. But when the song was done, he lowered his head and began again to graze. He had no quarrel with us; we were free to pass. We quietly stepped along the path toward him, turned the corner, and slowly walked toward the cabin.

But we broke and ran the last few yards. Close beside each other on the porch, looking through hemlocks to the broad brown back of the bear, safe, grateful, we all began to cry. So it was a bit tough to sing to the bear, but that's what we did, all the songs piled on each other in the harmony we felt. This was our ferry toll, this was the gift we gave in return for safe passage.

This was coming home, I remember thinking, not just to the cabin but to the beginning of a quieter, more trusting and respectful, reciprocal relation with the natural world.

In *Braiding Sweetgrass*, Robin wrote, "It has been said that people of the modern world suffer a great sadness, a 'species loneliness'—estrangement from the rest of Creation. We have built this isolation with our fear, with our arrogance, and with our homes brightly lit against the night. For a moment . . . those barriers dissolved and we began to relieve the loneliness and know each other once again."

I would like to be that bear, with his calm strength. I would be that bear, with the grace to accept a gift and give one in return.

III. Gates of the Arctic

WE KNEW BETTER THAN TO CROSS THE RIVER. ON THE FAR SIDE of the river from our camp, an Alaskan brown bear had been fishing all morning. He lay at the edge of a sort of cliff, about ten feet above the water, with his arms folded under his chin, staring into the pool below him. Without warning, he would launch himself off the cliff and belly-flop onto the water. From all the ruckus, the great wings of water, he would emerge every time with a big salmon in his jaws. He dragged the salmon up to the launch pad, gnawed on it absentmindedly, and then went back to fishing. He must have had a big pile of salmon up there, with just the brains bitten out. This was great fun to watch, hilarious even, anticipating the launch of the gigantic, shaggy mass and the thunderous splash. We were on the far side of the river, after all, and there were twelve of us and only one bear.

Things got serious in a hurry when a smaller bear wandered into view, heading up the beach toward the bear. The great bear sensed her before she sensed him, and we held our breath, half wanting to yell out a warning. The great bear sprang to his feet and, with astonishing speed, galloped toward the intruder. The chase was on, the smaller bear running for her very life. We expected the great bear would lose interest after a time, but he did not. He must have chased the other bear for at least a mile up the beach before he turned and came back to his pile of fish.

So we knew we had a defensive bear, and we knew he was defending a large swath of beach. To this day, I do not know why we accepted the guide's invitation to cross the river in canoes and hike up to fish a lake behind the first hill. I'm sure now that the bear was watching us from the moment we ground onto the beach, but we didn't see him until we were a quarter mile toward the lake. He was standing in the scrubby trees, looking hard at us. The direct gaze of a bear is not a good thing. The four of us grouped up and began to back away. The bear disappeared into the willows, then reappeared even closer. He stepped forward, raised up on his hind legs, dropped to all fours, galloped a few more paces toward us, and stopped. We drew every defensive armament we had—air horns, flares, bear spray—and stood there in a row like desperadoes ready for a gunfight. It's a wonder the bear didn't fall down laughing. He did not. He stepped closer. We stepped back. He stepped closer. We stepped back. Our fingers twitched over the various triggers, and all the nozzles pointed straight at the bear. He stepped closer. When we stepped back again, we found we were in a sort of ravine, out of view of the bear. We took off then, crashing through alders, splashing through streams, heading for the canoes. Those paddles dug deep and we were back in camp.

The bear appeared on the far shore, still staring us down, then returned to his food stash and recommenced flopping into the river. Frank discovered he had lost the tip of his flyrod but didn't go back for it. We ate lunch. I cried a little, of course.

I would like to be as fierce as that bear, defending my right to what sustains me.

IV. The Clark Fork, Montana

BEARS ARE SUFFERING FROM GLOBAL WARMING JUST LIKE OTHER species. In our cove, the winters are so warm that the snow melts in the bears' winter dens, soaking the bears. When they are forced from hibernation midwinter, they wander through leafless forests and puddles; there is nothing to eat except maybe the carcass of a starved deer. The cubs are born during hibernation, tiny little things that need to nurse, safe and warm, until spring. When the sows are forced from the dens, the cubs die. Desperate, the bears break into hunters' cabins, where they are shot dead. Defense of Life and Property. Perfectly legal.

In the summer, when rains don't come to fill the streams, salmon can't get over the gravel bars to spawn. Some of the salmon die in deep water and the bears starve. Other salmon stray to larger streams, where they congregate in great numbers in the overheated water. The big male bears gather there, defending the scarce food source against the females and yearlings, who either starve or lurk around human food sources and are shot dead. In severe weather—heat and drought, or freezing weather without the cover of snow—berry crops fail. Bears come down from the mountain meadows to forage in the intertidal zones, where they are shot by yacht-based trophy hunters. Here's a testimonial from one of them:

> "It's a big boar and a sow!" My heart started beating faster, as we realized we were looking at the spectacle of two very large brown bears making love! I got the brown bear of my dreams ... What a perfect bear he was, at 8'6".

Good lord.

So. I was leaving a session of the Society of Environmental Journalists, having just heard a representative of the American Petroleum Institute try to blame the disastrous Deep Horizon oil gusher on weather or bad concrete or some damn thing, anything but oil industry negligence. In the Q and A, he was just as eager to absolve fossil fuel companies from any blame for climate change. "We have met the enemy and it is us," he said, quoting poor little cartoon Pogo, an opossum, for god's sake. "If people didn't want oil, they wouldn't buy it, and if they didn't buy it, oil companies wouldn't drill," completely ignoring the decades and billions they have spent making sure that people are completely dependent on fossil fuels—killing alternative means of transportation, artificially lowering the price of fuel, undercutting renewable energy, and lying about it all to get money to drill some more.

As it happened, the Petroleum Institute guy and I left the room at the same time. Barely avoiding a collision, we walked out the door together. He was an enormous man, broad-shouldered in a long black overcoat. Shaved head, huge shoes. Hollywood could not have typecast a more appropriate industrial villain.

We were walking on ice. This is only half a metaphor. It's true that the bridge over the river was coated with ice, but it's also true that I was afraid of him and so I was thinking carefully about what to say. I had to engage him in conversation. This was too weird. I wanted to probe him for some softness in his opposition to any steps to curb global warming.

"So," I said. "Do you have children?"

It was sort of a sly approach to a discussion about the future, and I was congratulating myself, but he did not take the bait. He stopped walking and turned on me. Towering there, he pointed a huge finger at my face and said, "Don't you ever. Ever. Ever. Ever. Ever. Ever underestimate the strength of the fossil fuel industry."

Then he turned and walked away, and that was that. I should have yelled after him, "Fine. And don't you ever, ever, *la-de-da* ever underestimate the strength of mothers defending the lives of their children." But I didn't. Of course, I didn't. Sure, I cried just a little.

I would like to be a bear. To the extent that I am a bear, I pledge to find

ways to bring cunning and courage to the fight to stop the planetary plunder
that threatens the young ones of all species.

> The American Petroleum Institute is the best-funded lobbying
> force for climate policy in the United States. It spends $65 mil-
> lion a year, with a significant portion provided by ExxonMobil
> and Shell. "Extrapolated over the entire fossil fuel industry . . .
> it is not hard to consider that the [fossil-fuel industry] climate
> policy lobbying spending may be in the order of $500 million
> annually."
>
> (350.org, EcoWatch, Influence Map)

Hope Is Not the Thing with Feathers

A HISTORY

CRAZIEST THING. LOTS OF TIMES, WHEN AN ORGANIZATION ASKS me to talk about the climate crisis, the global extinction crisis, the crisis of ecosystem collapse, the diminishing prospects of the children, their request is very specific. "Nothing grim. People don't want to hear grim. Nothing angry or alarmist. Nothing too dark. Give us reasons to hope." Well, for crying out loud. In a power-hungry, profit-driven frenzy of extraction and pollution that can only be described as a moral monstrosity on a cosmic scale, international mega-corporations in collusion with thuggish petro-states are turning Earth into Venus, a planet that—it must be noted—does not support life. I guess this is what they don't want me to talk about. But under these conditions, what am I supposed to say?

A gentle request recently came from the Xerces Society, an admirable group of people who are passionate about invertebrates: butterflies of course, but also bees, worms, etc. "Can you give us *anything* to encourage us?" These are the people who last year watched the monarch butterfly populations in California fall by 86 percent. They cite British studies showing that thirty-six of the thirty-nine dragonfly species are declining. In the United States, local populations of fireflies are blinking out; worldwide, their numbers are in decline. The scientists' hearts must be broken. Unable to imagine how to cheer these people up without lying, I decided to take the academic's way out and write a brief history of hope.

66,000,000 B.C.E. THE OLDEST KNOWN FOSSIL RECORDS OF HOPE are found in the stomachs of dinosaurs who apparently continued to eat during their last few minutes on this planet, even as the skies lit up from a firestorm ignited by the impact of an asteroid that would wipe out 80 percent of them. That stubborn chewing is the definition of hope, I would say.

411 B.C.E. PERHAPS THINKING OF DINOSAURS, BUT MORE LIKELY thinking of his lieutenants, the great Athenian general Thucydides had a dim view of hope, noting that hopeful people "typically have a poor understanding of their situation, fail to come up with good plans, and things go badly for them in war."

A.D. 1891. MANY YEARS WENT BY. THEN THE MOST FAMOUS POEM ever written about hope was posthumously published. New England. Emily Dickinson.

> Hope is the thing with feathers
> That perches in the soul,
> And sings the tune without the words,
> And never stops at all.

This was a paradigm changer. Henceforth, hope was defined as something that soars in unbidden, out of the blue, not something we create ourselves. It does for us what music does, which is lift our spirits.

A.D. 1914. THE LAST PASSENGER PIGEON DIED IN CAPTIVITY AT the age of twenty-nine of an "apoplectic stroke," which made her tremble uncontrollably. That was the end of hope for that feathered thing.

A.D. 1942. HERE COMES ALBERT CAMUS, WHO THOROUGHLY DISapproved of hope. After all the ills of the world flew from Pandora's box, he said, the last and most dreadful horror came fluttering out. That was hope. Hope is a kind of resignation, he argued, and so it empowers all the other nastiness of the world by encouraging us to opt out of resisting. Why resist, when we can deceive ourselves into thinking that all will somehow be well? Hope, he went on to say, is like the painted screens that condemned prisoners held in front of their faces as they marched to their deaths, preferring a painting of swallows to a vision of the gallows. Forswearing hope, Camus died in an automobile accident, having written, "I know nothing more stupid than to die in an automobile accident."

A.D. 2011. AS AWARENESS OF CLIMATE CHANGE PICKED UP STEAM, novelist Barbara Kingsolver wrote, "If you run out of hope at the end of the day, get up in the morning and put it on with your shoes." She thus managed to write about hope without mentioning anything with wings (but read her great book *Flight Behavior*). She also contradicts Mr. Camus (and endorses the view of the Xerces Society) by implying quite straightforwardly that people need some hope of success if they are to continue the good work of making change.

A.D. 2012. A YEAR LATER, ECO-PHILOSOPHER JOANNA MACY taught that we actively choose hope and cultivate it in ourselves. Hope is something we do, not something we have (or can order online). Active hope is a decision each of us makes among three possible future scenarios:

1. Business as usual, striding toward the edge of the precipice until we tumble off. This choice is morally impossible.
2. The Great Unravelling, dwelling in despair as everything falls apart. This is psychologically impossible.
3. And the Great Turning, which she defines as "awakening to the dis-ease of our planet, our love of life and the revolution that can heal our world."

Active hope chooses the Great Turning as its work: to find the strength (or, if it is hard to find, to build up the muscles) to be an active participant in the work of creating a thriving future.

A.D. 2017. THE UNITED STATES ANNOUNCED ITS INTENTION TO withdraw from the Paris Climate Accords, causing millions of people to hope a bird would perch in the soul of the U.S. president and poop.

A.D. 2017. THE WHOLE CONCEPT OF HOPE BEGAN TO GROW A very tough skin, as time for meaningful change got shorter and odds got longer. Activist author Rebecca Solnit wrote (in 2005, but I first read it in 2017), "Hope isn't like a lottery ticket that you sit on the couch and clutch, feeling lucky. Hope is like one of those red axes that you pull out of its glass case, to break down doors in an emergency." Since this is definitely the Time of Emergencies, it is definitely also the Time of Axes.

A.D. 2018. THE WORLD'S LEADING CLIMATE SCIENTISTS WARNED that the world must cut its greenhouse gas emissions in half within a dozen years, or face significantly increased risks of "drought, floods, extreme heat and poverty for hundreds of millions of people." Hope plunged.

A.D. 2019. UNDER THE CIRCUMSTANCES, I WASN'T MAKING MUCH progress on writing my talk for the discouraged defenders of vanishing invertebrates. So, at a gathering of the Council for an Uncertain Human Future by the shore of a New England lake, I asked for advice. Here's what my friend Susanne Moser said: "Hope is when you look straight at the difficulties, and then you reach down deep inside you, to the very core of your wholeness, to find there the motivation to pick yourself up and keep on with this work of making meaning."

I thought that was exactly right, every clause of it. Hope is motivation to keep striving toward the most important possibility of the human condition, which is to create meaning from a world that presents itself as meaningless. And that motivation comes from inside you, from your moral wholeness, from the unity of belief and action. No birds, but courage and continuing.

I was still thinking about hope the next morning when I slid on wet grass and fell down a hill, breaking my ankle in three places. As emergency room doctors encased my ankle in plaster, I lay on my back and texted the organizers of the Xerces Society: "Broke ankle. Cannot talk to your group about hope."

A.D. 2019. THE XERCES SOCIETY CONVINCED LEPIDOPTERIST Robert Michael Pyle to step into the program in my stead. With good will and stalwart courage, he read poems that celebrate the gloriously blue mariposa butterfly which, being scaled, has nonetheless learned to fly without feathers. So may it be with hope.

A.D. 2020. REALIZING THAT THE FUTURE IS SHAPED BY THE STO- ries we tell each other, news outlets have begun publishing good news, including "99 Good News Stories You Probably Didn't Hear About." Pop- ulations of South Atlantic humpback whales have reached 93 percent of the numbers they had reached before they were hunted almost to extinction. China is creating the largest national park in the world to protect giant pan- das. Ecuadorian courts protected a huge chunk of Indigenous land holdings from mega-oil corporations. A wildlife reserve in Mozambique protected every single elephant from poachers for a full year. South Africa increased its proportion of protected waters from 0.4 percent to 5.4 percent. The United States passed a new law protecting animals from some forms of cruelty. And ninety-four other stories.

"Game on," I say to the industrial whalers, panda-bear habitat loggers, rainforest-destroying mega-oil corporations, elephant poachers, industrial- strength fishing fleets, Ethiopian loggers, and cruel Americans. The struggle for the future of the Earth is now officially engaged.

But note: Every one of these good news stories is a story not so much about hope, but about struggle, about brawls to protect something of last- ing value—life on Earth, for example—from those who would destroy it for short-term profit or pleasure. This will be a struggle. Make no mistake. We can tell how close civilization is to making the Great Turning by the fury of the forces arrayed against change. We should not be caught un- awares. The time of any paradigm change is a dangerous period of shouting and bullies.

Call in all the feathered things that are perched somewhere in your weary soul. Call in the harpy eagles and the sharp-shinned hawks. Call in the booming cassowaries and the shrikes. Call in whatever character traits *Velociraptor* and extravagantly feathered *Tyrannosaurus rex* have embedded in your reptilian brain. Hope is not all we need. What we need is strength— strength in numbers and strength in moral conviction. What we need is shrieking, roaring courage.

The 2018 Western Monarch Thanksgiving Count found that the number of West Coast monarch butterflies spending the winter in California had plunged to only 20,456 individuals—a drop of 86 percent since the previous year. In comparison, the asteroid that slammed into the Earth 66 million years ago wiped out "only" 70 to 80 percent of life on Earth.

(*National Geographic*)

Why We Won't Quit

LAST MONTH, ON THE WAY FROM ONE CLIMATE MEETING TO AN-other, my friend SueEllen Campbell and I stopped along the Oregon coast to camp for the night. Leaving our husbands to light the lantern and pitch the tents, we walked down to the beach to watch a red sun set through purple clouds. While parents gathered up their families, lingering children stood ankle-deep in pink water, looking out to sea. A flock of gulls flew north. For a long time, neither of us said anything. It was enough to take it in, the comradery of standing shoulder to shoulder in the peace of that moment.

Out of the blue, SueEllen asked, "Why do you suppose we keep doing this work?" Of course, we laughed. We are old veterans of climate-change and extinction struggles who have tried to do our part, in every way we know how, to keep our fossil fuel–addicted civilization from driving off a cliff and taking half the plants and animals with it.

Are we tired? Sure. Discouraged? Absolutely. Pissed off? Yep. Sad? Call it brokenhearted. Quitting?

It might be time. Game over, friends and experts tell us; we're doomed. It's true that the news about the linked terrors of global warming, species extinction, and ecocultural collapse is awful. More and bigger wildfires, great swaths of drought and starvation, stronger and wetter hurricanes, floods of all kinds, coastal villages a few storms away from destruction, feedback loops kicking in as methane leaks from melting tundra and heat-absorbing soils replace reflective ice, hundreds of thousands of refugees seeking safety as the world turns cruel, institutions of truth and order breaking down. Destructive ways of living are skillfully protected by tangles of profit and power around the globe, and we are running out of time. The Intergovernmental Panel on Climate Change now gives the world ten years to cut global greenhouse gas emissions in half, if we are to stop warming at "only" 1.5°C and extinctions at "only" 40 percent. Don't think quitting hasn't crossed our minds.

But maybe to our surprise, reasons why we can't quit poured out of us, one good reason after another. We sat down on a log. This was going to be a conversation we would want to remember. I pulled out my journal to get it down.

"We are doing this work," SueEllen said, "because we are *not* doomed, as long as we act. A world where we do everything we can to restrain climate change barely resembles one where we do nothing. We won't like the first world, but we might not survive in the second."

"Yes," I added, "and because I want to be the kind of person who doesn't give up on important jobs. You don't do what's right because you think it might get you something. You do it because it's right. That's what integrity is—doing what you believe in, even if it won't save the world."

SueEllen paused for a long moment. "And because I won't walk away from the hurting world any more than I will walk away from my mother as she grows old and frail and sometimes confused," she said. "I love her and owe her and have a duty to her and admire her and enjoy her company."

"Yes, and because I promised my newborn children: 'I will always love you. I will keep you safe. I will give you the world.' I didn't mean, 'I will give you whatever is left scattered and torn on the table after the great cosmic going-out-of-business sale.' I meant, 'I will give you this beautiful, life-sustaining, bird-graced world.'"

Truth be told, it just about kills me to think about the broken promises to the children. And when I think of my mother—if she learned that her daughter's generation was losing the battle for the beloved Earth's future, she would leap from her grave to struggle alongside us.

The wind was coming up as the sun went down. We could see the rough seas etched against the red disk of the sun. The seas were gray now, flecked with white. The bell-buoy at the Yaquina River bar began to rock and moan. Only a single family was left on the beach, and they were hurrying their children up the dune.

"Everybody knows what we have to do," SueEllen said. "It isn't as though the world is waiting for some technological breakthrough or divine revelation. We just need to stop setting carbon on fire."

More and more reasons poured out of us. I scratched madly to keep up. We clearly had been having this conversation with ourselves for years, and it was a relief to share our reasons, to spill them into the salty air.

"We have to do this work," I said, "because climate change is unjust. It threatens the greatest violation of human rights the world has ever seen. But injustice is cowardly and fragile; it crumbles when people stand up for what is right."

"Yes," SueEllen said. "And we don't want to be free riders, taking advantage of the actions, often sacrifices, of the people who step up. If we avoid planetary ruin, if we find better ways to live, it will be because of the courage of those who act."

"Yes," I agreed and paused. I was struggling not to cry. "Sometimes I think that I keep going because I am wearing my dad's rubber boots. They are too big for me, but my own are old and leaky. So I am walking in the boots he wore at the edge of all the marshes he defended until the day he died. If you are walking in the shoes of a hero, you can't exactly turn back."

SueEllen waited to give my pen time to catch up. "And, Kathy," she said, "we can't and therefore don't have to solve the whole problem alone. We only have to help where and how we can. So many good people are in this fight with us, in governments around the world, in businesses, in states and towns and neighborhoods and churches. They are smart and experienced and empowered by a vision of a planet redeemed. I believe, and choose to believe, that in this emergency, as in every emergency, more of us will come out to help each other than will rush in to exploit and loot."

"Yes. Exactly. We do this work because despair is lonely and useless while climate action is full of friendship, satisfaction, and glee. You get to hang out with people who care as much as you do and act with remorseless resolve. Taking action is the only real cure for hopelessness. It feels good, and important, like you're not wasting your life on small things."

"We have so much to lose and so much left to save—everything from birdsong to our own sorry souls."

We stayed at the beach until the stars came out and the breeze came up. Still mulling over our reasons to stay in the struggle, we walked back to camp on a mossy trail through a tunnel of spruce trees. A Swainson's thrush sang and would not stop singing, even in the deepest dusk, and that was another reason. The deep moss was a reason. So were the ancient trees. So were the children wading in the swash.

In early 2020, about seven in ten Americans (72 percent) think global warming is happening, while about one in ten (10 percent) think it is not happening. About six in ten Americans (62 percent) say they are "somewhat worried" about global warming. One in five (23 percent) are "very worried" about it. Very few Americans (12 percent) think it is too late to do anything about global warming.

(Yale Program on Climate Change Communication)

Epilogue

Sing Out from the Mountaintops

ONE OF THE MANY THEORIES OF THE ORIGIN OF HUMAN SINGing puts people alone in a wild place, urgently needing to call for help. What do they do? They could speak into the empty clearing: 50 decibels. They could shout: 88 decibels. Or they could sing out. An opera singer can produce 100 decibels of sound, more than twice as loud as shouting. A yodeler can produce 105. So when I try to imagine the origin of human song, I picture a young woman wrapped in animal skins, carrying a child. She is scrambling to the top of a crag to escape a cave bear that has been stalking her. She lifts her head and sings clear tones that can be heard miles away, through a forest, maybe, or across a river. From all around the forest, family members turn and stride toward her, fully alert.

In whatever way the skill evolved, in whatever voice, humans have the power to call to the community for help in hard and dangerous times. If there ever was a time when we need to sing out, it is now. The forces arrayed against the planet's living beings are formidable, the dangers existential.

The major causes of the extinction crisis? Habitat loss through deforestation, mining, urbanization, and other destructions. Global warming.

Spread of exotic species. Overharvesting, overhunting, overexploitations in seemingly infinite variety. Pollution and poisoning. All these are exacerbated in every way by human population growth. These are not cosmic forces, uncontrollable; these are the results of human decisions. They are the unsurprising consequences of an extractive capitalist system engorged by its own excesses. Any culture that prides itself on accumulating wealth instead of sharing it, any culture that gobbles up the fecundity of the planet instead of nurturing it, any economy that eats its own feet, any economy of infinite extraction, will kill off the sources of its material and spiritual sustenance—the growing things, plants and animals—and ultimately itself.

Now the world finds itself at an intermezzo. The whole world is extemporizing, wildly searching through the cruel and calamitous chaos of our time for a path toward a culture that is lasting, beautiful, and somehow redemptive. It is a time to let our imaginations soar, not to envision the end of the world, but to set a compass course for its reinvention. Even if we have to change our lives forever, there is still a chance to save what we love too much to lose.

I am convinced that people love the Earth's wild music, the harmonies that comfort and delight them, inviting them out of their narrow preoccupations and welcoming them into the raucous kinship of the family of life. I am convinced that people want to defend the wild family. If they didn't know before how important natural places are to their well-being, they know now, after the COVID-19 virus lockdowns kept them indoors.

How do we save Earth's wild songs and the creatures who sing them?

Some people might argue that it's too late. But what does that mean? Too late for what? Too late for the golden toad? Almost certainly. Too late for monarch butterflies? Probably. Too late for civilization as we know it? I'd say, yes, unless something changes fast. Too late for human beings? I don't know yet, but I would be surprised, given the resilience of the species. "It's not the strongest of the species that survives," Darwin wrote, "nor the most intelligent, but the one most responsive to change."

The point is, we are not in a game that ends when a buzzer sends the losing team, dejected, off the court forever. There's a long way to go before it's too late for life, large and small; life will go on for probably 5 billion years, until the sun expands and swallows the Earth in fire. What we do now will change everything forever. Whatever species get through the narrow hourglass of this century will determine what lives will evolve on the planet through that future. If you don't believe it, think how consequential it turned out to be that small mammals were able to save themselves in dank tunnels as the blast from the asteroid scorched the planet in the Fifth Great Extinction and took out the dinosaurs.

Because it's not too late, we can't give up and lie supine before the human steamroller of destruction. We weren't around to stop the microbes, volcanoes, and asteroids that caused the Third Great Extinction that killed the trilobites. We weren't around to plug the volcanoes that caused the Fourth Great Extinction that killed the archosaurs and large amphibians. We couldn't send rockets to blow up the asteroid that caused the Fifth Great Extinction that killed the dinosaurs. But we are in a strong position to stop the Sixth Extinction. Our greatest asset in that struggle, maybe our only hope, is the beauty and power of our voices, raised in a global chorus of outrage, conscience, and imagination.

What does that chorus call us to do?

Three things, and we have to do them all: Stop the killing. Defend everything that is left. Create new lifeways in harmony with the Earth.

Number one: We will stop this mad killing. It's true that thousands of species are irretrievably gone. It's true that regulatory agencies that might have outlawed poisons or defended clean air, clean rivers, and migratory birds have been captured or purchased outright by the very corporations they were meant to regulate. So corporations are free to make a profit and damn the consequences, particularly if those consequences are borne by those who have no voices to defend themselves—small children, plants and animals, marginalized people, future generations of all species, those Pope

Francis calls "the silent voices screaming up to heaven." It's true that change, if change occurs, will be led—not from some sudden moral awakening in the government, but from the conscience of the streets. That is you and that is me.

We can stand against corporate wreck and plunder. Stand. In the way. With a choir and a conscience and a sign. We can stand and say, "This is wrong, and I will not be a part of it."

That's good. We also can stand and say, "This is wrong, and I will not *allow* it."

Together, we can draw the line, calling out, "Not another mountaintop. Not another rainforest. Not another estuary. Not another prairie. Not another mighty river can be traded away for cash. These are not industries' to take or sell. They belong to the future of the everlasting Earth."

Now, number two: Ferociously defend what we have left—all the urgent, innocent lives. Noah knew that whatever survived the Great Flood would repopulate the world, the lions and the elephants two by two. Now, millennia later, the world is going through a biological bottleneck as brutal as God's fury in Noah's time. Whatever species make it through—that's what the world will be made of. Noah protested. "I'm old, I'm tired, why me, o Lord?" The answer is, it's got to be everybody, each asking, what ark can I build, what habitat can I save or create or restore, that will carry living things?

By now, Earth's plants and animals are refugees, driven from their hedgerows and homes by rising water and war, crowding at the shore, looking desperately for safe harbor. No less than humans, they deserve a place to put down roots, to spread their wings, to raise their young. The world needs flotillas of arks, rescue boats uncountable, says poet David Oates. "Tiny handmade parks and massive prairies, ocean reserves bigger than aircraft carriers, lovingly restored arks like marshlands, all set into the forward-river of time, to carry the creatures, great and small, through the rocky shoals of our era. . . . and then to touch, chancewise, on dry land. And start the world again."

Number three, the third and necessary thing? Create lifeways of respect and restraint that bring civilization into harmony with Earth's swirling music. There has got to be a way to live on a beautiful planet without wrecking it. There has got to be a better way. Our work is not to save our way of life, but to save the world from this way of life's destructive power. The cosmic challenge of our time is to re-create our humanity through this great crisis, and this means realizing our full humanity as it evolves in kinship with all the world's blooming, bellowing lives—so that we can learn what it means to live in concert with the Earth. This will require moral courage and a clear vision of a human civilization worthy of the Earth's wild music.

> On the reeling planet that we hold in our hands,
> may gentle rain fall forever on green hills,
> may ice come in its time to glaze the bays.
> May salmon faithfully return when sandpipers call.
> May songbirds sing in the apple trees.
> And may the children hum themselves to sleep in a safe and sustaining
> night.

Acknowledgments

A CHORUS OF FRIENDS

A WRITER SOMETIMES FEELS THAT SHE WRITES ALONE, A SOLOIST with a thin, wavering voice, searching uncertainly for the right tone. The truth is, she is only one part of a chorus of friends and literary forebears who bring their voices to the song. What an extraordinary gift this is, to be part of the choir.

So thank you to the gifted friends who read *Earth's Wild Music* in manuscript form and made helpful, sometimes transformative, responses. They are Frank Moore, Charles Goodrich, Carol Mason, and George Mason.

And thank you to so many other smart friends, relatives, audiences, and witting and unwitting strangers who shared ideas and challenged me to think more deeply and write more carefully about them. Primary among these is my beloved Frank, who knows or can find answers to any question I ask him, whether I whisper a question in bed, yell down from my upstairs office, or interrogate him on the trail. Other brilliant friends and trail-mates are Sarah Buie, John Calderazzo, SueEllen Campbell, Alison Hawthorne Deming, Marty Fisk, Nara Garber, Bob Haverluck, Gordon Hempton, Robin Wall Kimmerer, Carolyn Kremers, Hank Lentfer, Susanne Moser,

Rachelle McCabe, Erin Moore, John Moore, Jonathan Moore, Zoey Moore, the late Richard Nelson, Janet Nielsen, Hob Osterlund, Peter Raven, Kim Rivera, Libby Roderick, Kim Stafford, Fred Swanson, Mary Evelyn Tucker, Julianne Warren, Gail Wells, Clement White-Moore, and Theodore White-Moore. There is no greater thrill than to listen to any of them think out loud.

I am grateful to organizations that have fostered my thinking, through invitations to speak and write or by providing a quiet place to ponder. These include the Center for Humans and Nature; the Council for the Uncertain Human Future; Friends of OSU Old Growth; Long-Term Ecological Reflections at the H. J. Andrews Research Forest; Mesa Refuge; the Spring Creek Project for Ideas, Nature, and the Written Word; the United States Forest Service; and the Xerces Society.

Warm thanks go to Laura Blake Peterson, of Curtis Brown, and Jack Shoemaker, Jennifer Alton, and all the folks at Counterpoint, for loyal advocacy and respectful guidance.

Finally and foremost, I give thanks to waving grasses, wind in cottonwoods, whimbrels, killer whales, crocodiles, and all the other members of the choir.

Notes

A version of "The Sound of Human Longing" originally appeared in *Pine Island Paradox* by Kathleen Dean Moore (Minneapolis: Milkweed Editions, 2004). Copyright © 2004 by Kathleen Dean Moore. Reprinted with permission from Milkweed Editions. milkweed.org.

"Repeat the Sounding Joy" first appeared in *Wild Comfort* by Kathleen Dean Moore © 2010. Reprinted by arrangement with Shambhala Puiblications, Inc., Boulder, Colorado. www.shambhala.com.

A version of "Songs in the Night" originally appeared in *Pine Island Paradox* by Kathleen Dean Moore (Minneapolis: Milkweed Editions, 2004). Copyright © 2004 by Kathleen Dean Moore. Reprinted with permission from Milkweed Editions. milkweed.org.

"Listening for Bears" is a revision of my article "Bear Sign (On Joyous Attention)," published in Tom Fleischner, ed., *The Way of Natural History* (Trinity University Press).

"The Silence of the Humpback Whale" can also be found online in the *Ecological Citizen*.

"The Angelus," an essay included in "The Meadowlark's Broken Song," was first published in abbreviated form in *Moral Ground*, edited by Kathleen Dean Moore and Michael P. Nelson © 2010 (Trinity University Press).

In "The Terrible Silence of the Empty Sky (Seashore)," the ghost-net scene is adapted from a passage in my novel, *Piano Tide* (Counterpoint).

"Twelve Heartbreaking Sounds That Will Remain" was first published in *Orion*.

"Living Like Birds" was first printed in French translation in *America*.

A version of "Late at Night, Listening" originally appeared in *Pine Island Paradox* by Kathleen Dean Moore (Minneapolis: Milkweed Editions, 2004). Copyright © 2004 by Kathleen Dean Moore. Reprinted with permission from Milkweed Editions. milkweed.org.

"Silence Like Scouring Sand" was first printed in somewhat different form in *Orion*.

"The Song of the Canyon Wren" is reprinted from *Holdfast* by Kathleen Dean Moore © 1999. Reprinted by arrangement with Oregon State University Press, Corvallis, Oregon. www.osupress.oregonstate.edu.

"How Can I Keep from Singing?" is adapted from my article of the same name, published in Goodrich, Moore, and Swanson, eds., *In the Blast Zone: Catastrophe and Renewal on Mount St. Helens* © 2008. Reprinted by arrangement with Oregon State University Press.

A version of "After the Fire, Silence and a Raven" originally appeared in *Pine Island Paradox* by Kathleen Dean Moore (Minneapolis: Milkweed Editions, 2004). Copyright © 2004 by Kathleen Dean Moore. Reprinted with permission from Milkweed Editions. milkweed.org.

Portions of "We Will Emerge Full-Throated from the Dark Shelter of Our Despair (The Dawn Chorus)" can be found online in Terrain.org's series "Dear America."

"Rachel's Wood Pewee (On Wonder)" is a revised and expanded version of "The Truth of the Barnacles: Rachel Carson and the Moral Significance of Wonder," in *Rachel Carson: Legacy and Challenge*, coedited by Lisa H. Sideris and Kathleen Dean Moore and published by the State University of New York in 2008.

"Why We Won't Quit" is a variation of an essay that appeared online in *Earth Island Journal*. Special thanks to its coauthor, SueEllen Campbell.

Portions of the Epilogue were first printed in *Great Tide Rising* by Kathleen Dean Moore © 2016. Reprinted by arrangement with Counterpoint Press, Berkeley, California.

© Frank Moore

KATHLEEN DEAN MOORE is the author or coeditor of many books about our moral and emotional bonds to the wild, reeling world, including *Wild Comfort*, *Moral Ground*, and *Great Tide Rising*. She is the recipient of the Pacific Northwest Booksellers Association Book Award and the Oregon Book Award, along with the WILLA Literary Award for her novel *Piano Tide*. A philosopher and activist, Moore writes from Corvallis, Oregon, and Chichagof Island, Alaska. Find out more at riverwalking.com.